D1165652

Guide to the Deterioration and Failure
of Building Materials

Guide to the Deterioration and Failure of Building Materials

R. O. Heckroodt BSc, MSc, DSc(Pret), Dip Ceram (Leeds)

Emeritus Associate Professor, University of Cape Town

 ThomasTelford

Published by Thomas Telford Publishing, Thomas Telford Ltd, 1 Heron Quay, London E14 4JD.

URL: http://www.thomastelford.com

Distributors for Thomas Telford books are
USA: ASCE Press, 1801 Alexander Bell Drive, Reston, VA 20191–4400, USA
Japan: Maruzen Co. Ltd, Book Department, 3–10 Nihonbashi 2-chome, Chuo-ku, Tokyo 103
Australia: DA Books and Journals, 648 Whitehorse Road, Mitcham 3132, Victoria

First published 2002

Also available from Thomas Telford Books

Deteriorated concrete: inspection and physicochemical analysis, F. Rendell, R. Jauberthie and M. Graham. ISBN 07277 3199 X
Concrete reinforcement corrosion: from assessment to repair decisions, P. Pullar-Strecker. ISBN 07277 3182 3
Portland cement: composition, production and properties, 2nd edition, G. C. Bye. ISBN 07277 2766 4

A catalogue record for this book is available from the British Library

ISBN: 0 7277 3172 6

Typeset by Helius, Brighton and Rochester
Printed and bound in Great Britain by MPG Books, Bodmin, Cornwall

Preface

Professionals concerned with the built environment are all too often confronted with cases where building materials have failed prematurely. The information required for the understanding of the causes of such failures, or for the appropriate remedial action is available in a number of excellent texts, but it is generally buried under a mass of other information. Numerous trade bulletins are also available, but they usually only deal with a particular group of materials.

This guide condenses and organises the relevant information that should allow the professional to take appropriate action, or at least understand the recommendations made by specialists. Although the assumption is that the reader already has an understanding of the fundamental aspects of construction materials, references to further detailed data are provided.

The distinction between the deterioration and failure of building materials and system failures of structures is sometimes rather blurred. This guide is mainly concerned with material failures, while system failures are only referred to when the possibility of confusion exists.

R. O. Heckroodt
Hermanus, July 2002

Acknowledgements

My sincerest thanks are due to my wife Erma for her constant encouragement and patience during the preparation of this book.

I am greatly indebted to the following friends and colleagues for their valuable contributions to and reviews of sections of this guide:

J. E. Krüger, DSc(Pret), PrSciNat
Consultant: Building Materials Science and Technology
Previously: Head, Inorganic Materials Division, National Building Research Institute, SA CSIR

J. R. Mackechnie, MSc(Eng), PhD (UCT), PrEng, MSAICE
Cement and Concrete Association Fellow, University of Canterbury, New Zealand
Previously: Research Officer, Concrete Research, Department of Civil Engineering, University of Cape Town, Republic of South Africa

T. Rypstra, MSc, PhD (Stell), STD (Stell)
Senior Lecturer, Department of Wood Science, University of Stellenbosch, Republic of South Africa

M. S. Smit, BA (Pret), MPISA
Project Leader and Chief Research Officer, Division of Building Technology, SA CSIR

R. O. Heckroodt
BSc, MSc, DSc(Pret), Dip Ceram (Leeds)
Emeritus Associate Professor, University of Cape Town

Contents

Preface . . . . *i*

Acknowledgements . . . . *ii*

Introduction . . . . *v*

Chapter 1 Deterioration of concrete **1**
Corrosion damage to reinforced concrete 2
Alkali–aggregate reaction 22
Chemical attack on concrete 25
Fire damage 27
Frost attack 28
Dimensional change 29
Further reading 31

Chapter 2 Metal failure **33**
Modes of metal failure 34
Suppression of metal corrosion 37
Structural steel 38
Galvanised steel 42
Weathering steel 44
Stainless steel 44
Aluminium 46
Copper alloys 49
Further reading 50

Chapter 3 Deterioration of masonry **51**
Causes of deterioration 52
Cement-based mortar and plaster 60
Natural stone masonry 61
Calcium silicate masonry 66
Fired clay masonry and paving 67
Concrete bricks and blocks 73
Further reading 74

Chapter 4 Tiling failures 75
Material failure 76
System failure of tiled floors and walls 81
Further reading 85

Chapter 5 Deterioration of timber 87
Agents of attack 88
Remedial actions 99
Preventive actions 101
Further reading 104

Chapter 6 Paint failures 107
Paint failures 107
Causes of failure 108
Preventive and remedial actions 114
Further reading 115

Chapter 7 Examples of failure 117
Concrete failure 118
Metal failure 124
Deterioration of masonry 128
Tiling failure 138
Timber deterioration 142
Paint failure 150

Bibliography 157

Index 159

Introduction

First of all it must be recognised that all materials have a limited lifetime, but it is situations of premature failure that should cause concern. There are many reasons why defects develop prematurely in buildings. Unfortunately, poor practices (often through ignorance of well-recorded facts) and bad workmanship and supervision are all too frequently the reasons for failure. However, this guide is mainly concerned with material failures, and therefore system failures are only referred to when the possibility of confusion exists.

Interaction between environmental conditions, material properties and structural factors must constantly be considered when evaluating the failure of building structures. More often than not more than one disruptive mechanism may operate and a detailed analysis of the affected structure is imperative to identify all the causes that contributed to the deterioration of the material. For example, cracking of concrete structural members due to overloading could effectively decrease the cover depth, resulting in early corrosion of the reinforcement. Conversely, spalling of the cover concrete due to reinforcement corrosion decreases the load-bearing ability of structural members, possibly resulting in overload conditions. Fungal growth on timber requires high humidities that could be the result of poor ventilation, while inadequate water shedding could be the reason for localised frost damage. An accumulation of pollutants could be the underlying reason for metal corrosion, or paint failures could be ascribed to fluctuating moisture content in a timber substrate.

This guide provides a concise presentation of important information to make the reader aware of the causes of deterioration of building materials. It also briefly provides suggestions and recommendations on how best to deal with them. There are usually a number of possible repair actions and the task is to find the most appropriate solution, taking into account long-term as well as short-term costs, durability, effectiveness, feasibility and environmental acceptability. Obviously the repair should not perpetuate the problem.

Many agencies – physical, chemical and biological – originating from environmental impact are responsible for the deterioration of building materials. The central villain in most material failures is water, in vapour, liquid or solid form. It is the carrier of harmful contaminants, creates conditions for chemical processes and sustains biological actions. The role of moisture in the deterioration of building materials should never be underestimated, and the first step in the remedial process would usually be to rectify conditions that allowed the ingress of moisture into the building elements.

Chapter 1

Deterioration of concrete

Contents

Corrosion damage to reinforced concrete 2
Manifestations of corrosion damage 2
Condition surveys of reinforcement corrosion 7
    Visual assessments 7
    Chloride testing 7
    Carbonation depth 12
    Rebar potentials 12
    Resistivity 13
    Corrosion rate measurements 14
Repair strategies 14
    Patch repairs 15
    Migrating corrosion inhibitors 19
    Electrochemical techniques 19
    Cathodic protection systems 21
    Demolition or reconstruction 22
Alkali–aggregate reaction **22**
Types of alkali–aggregate reaction 22
    Alkali–silica reaction 22
    Alkali–carbonate rock reaction 22
Recognition of alkali–aggregate reaction 22
    Cracking of concrete 23
    Expansion of concrete members 23
    Presence of gel 23
    Discoloration 23
Confirmation of alkali–aggregate reaction 23
Conditions necessary for alkali–aggregate reaction 24
    High alkalinity of pore solution 24
    Reactive phases in the aggregate 24
    Environmental conditions 24
Preventive measures 24
Remedial action 25
    Reaction dormant 25
    Reaction active 25

Chemical attack on concrete **25**
Soft water 26
    Attack mechanism 26
    Remedial action 26
Sulphates 26
    Attack mechanism 26
    Remedial action 27
Fire damage **28**
Effect of high temperatures 28
Remedial action 28
Frost attack **28**
Damage mechanism 28
Remedial action 29
Dimensional change **29**
Shrinkage and creep 30
    Drying shrinkage 30
    Carbonation shrinkage 30
    Creep 30
Thermal expansion 30
Remedial action 31
Further reading **31**

The dominant cause for failure of concrete in structures is corrosion of the reinforcing steel. The other causes are less common, but still critical, agents of material failure. It is important to constantly bear in mind that the failure of concrete structures can seldom be ascribed either exclusively to the failure of a material component (cement, aggregate or reinforcement) or exclusively to failure of the system (structural or design failure). The reasons why concrete deteriorates are summarised in Table 1.1.

Corrosion damage to reinforced concrete

Manifestations of corrosion damage

Reinforcement corrosion is the major cause of deterioration of concrete structures. Corrosion results in a volume increase of the steel of up to ten times its original volume due to the formation of hydrated oxides. This expansion of the steel results in mechanical disruption of the encasing concrete.

Reinforcement corrosion is particularly pernicious in that damage may occur rapidly and repairs are invariably expensive. Furthermore, by the

Table 1.1 *Concrete deterioration diagnostics*

Visual appearance of deterioration	Type of deterioration and causes	Confirmatory testing
Large areas of rust stains, cracking along pattern of reinforcement, spalling of cover concrete, delamination of cover concrete	*Reinforcement corrosion*: exposure to normal climatic conditions, with cyclic wetting and drying	Cover depth of rebar Carbonation and chloride testing Exploratory coring Electrochemical testing
Expansive map cracking, restrained cracking following reinforcement, white silica gel at cracks	*Alkali–aggregate reaction*: concrete made with reactive aggregates	Core analysis for gel and rimming of aggregates Petrographic analysis Aggregate testing
Deep parallel cracking, pattern reflects reinforcement positions	*Drying shrinkage/creep*: initial too rapid drying, long-term wetting and drying cycles	Concrete core analysis Loading and structural analysis Aggregate and binder analysis
Deterioration of surface, salt deposits on surface, cracking caused by internal expansive reactions	*Chemical attack*: exposure to aggressive waters (e.g. domestic and industrial effluent)	Chemical analysis of concrete Core examination for depth of attack and internal distress
Surface leaching of concrete, exposed aggregate, no salt deposits	*Softwater attack*: exposure to moving fresh waters (slightly acidic) in conduits	Chemical analysis of water Core examination for leaching Aggregate and binder analysis
Surface discoloration, concrete spalling, buckling, loss of strength, microcracking	*Fire damage*: exposure to open fires sufficient to cause damage	Core examination for colour variations, steel condition Petrographic analysis Specialist techniques
Major cracking and localised crushing, excessive deformations and deflections of structural members	*Structural damage*: structure subjected to overload	Loading and structural analysis Core testing for compressive strength and elastic modulus

Table 1.2 *Conditions and features of reinforcement corrosion*

Type of corrosion	Environment or causative conditions	Significant features of deterioration
Chloride induced	Marine environments	Distinct intense anode and cathode regions
	Industrial chemicals	Rapid and severe localised pitting corrosion and damage to surrounding concrete
	Admixed chlorides (older structures)	Corrosion damage affecting structural integrity may be far advanced before being noticed (surface stains, cracking, spalling)
		More pernicious and difficult to treat than carbonation-induced corrosion
Carbonation induced	Unsaturated concrete	General corrosion with multiple pitting along rebars
	Polluted environments	Moderate corrosion rates, except when wet and dry faces are near to each other
	Low cover depths to steel	Corrosion damage easily noticed (surface stains, cracking, spalling); generally only affects aesthetics
		Requires a different repair approach from chloride-induced corrosion; repairs are generally successful
Stray current	DC power supplies	General corrosion of rebar exposed to moist conditions
	Railway systems	Corrosion not confined to low cover depths
	Heavy industry, smelters	Large crack widths possible

Category	Causes	Description
Chemical induced	Sulphate in groundwater; Fertiliser factories; Industrial plants; Sewage treatment works	Corrosion generally associated with near-saturated conditions; Concrete deterioration occurring together with corrosion
Secondary forms	Primary cracking due to alkali–aggregate reaction, delayed ettringite formation, structural cracking	Corrosion localised in regions where cracks intersect the rebar; Other forms of distress evident in concrete (i.e. alkali–aggregate reaction gel deposits)
Artificially induced	Bimetallic corrosion; Partial sealing of concrete; High temperatures (>200°C); Patch repairs of corrosion	Generally very localised intense corrosion due to well-defined anode/cathode regions

time visible corrosion damage is noticed, structural integrity may already be compromised. There are two major consequences of reinforcement corrosion:

- cracking and spalling of the cover concrete as a result of the formation of the corrosion products
- a reduction of cross-sectional area of the rebar by pitting (a problem in prestressed concrete structures).

See also:
    Metal, Corrosion, p. 34
    Metal, Corrosion of steel embedded in concrete, p. 39

Unfortunately, most reinforced concrete structures that exhibit cracking and spalling have gone beyond the point where simple, cost-effective measures can be taken to restore durability. *Condition surveys* are therefore an important strategy for identifying and quantifying the state of corrosion of a structure over time. The results of such surveys will determine the most appropriate *repair strategy*.

The features of reinforcement corrosion induced by different conditions are summarised in Table 1.2 and the factors that influence the manifestation of reinforcement corrosion are listed in Table 1.3. Figure 1.1 illustrates the three stages in the development of corrosion of reinforced concrete structures.

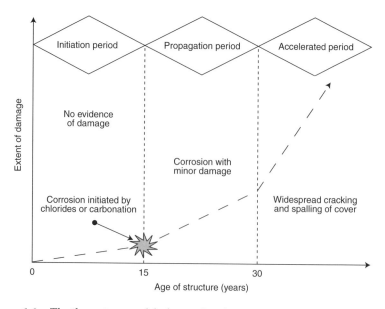

Figure 1.1   *The three-stage model of corrosion damage*

Table 1.3 *The manifestation of reinforcement corrosion*

Factor	Influence
Geometry of the element	Large-diameter bars at low covers allow easy spalling
Cover depth	Deep cover may prevent full oxidation of corrosion product
Moisture condition	Conductive electrolytes encourage well-defined macro-cells
Age of structure	Rust stains progress to cracking and spalling
Rebar spacing	Closely spaced bars encourage delamination
Crack distribution	Cracks may provide low resistance paths to the reinforcement
Service stresses	Corrosion may be accelerated in highly stressed zones
Quality of concrete	Severity of damage depends on the concrete quality

Condition surveys of reinforcement corrosion

A detailed corrosion or condition survey is vital in order to identify the exact cause and extent of deterioration, before repair options are considered. Unfortunately, most reinforced concrete structures that exhibit cracking and spalling have gone beyond the point where simple, cost-effective measures can be taken to restore durability. Condition surveys are therefore an important strategy for identifying and quantifying the state of corrosion of a structure over time. The results of such surveys will determine the most appropriate repair strategy. The various survey techniques are summarised in Table 1.4.

Visual assessments
Items that should be included in a checklist for a visual assessment of concrete degradation are listed in Table 1.5. Visual assessment of deterioration may come too late for cost-effective repairs because rebar corrosion damage often only fully manifests itself at the surface after significant deterioration has occurred.

Chloride testing
Chlorides exist in concrete as both bound and free ions, but only free chlorides directly affect corrosion. Measuring free water-soluble chlorides accurately is extremely difficult, and chlorides are therefore most

Table 1.4 *Condition surveys to evaluate reinforcement corrosion*

Surveys	Comments
Visual: use comprehensive checklist	Corrosion during early stages not visible
	Visual survey first action of detailed investigation
Delamination: hammer or chain drag	Often underestimate full extent of delamination and internal cracking
	Not definitive
Cover surveys: use alternating magnetic field to locate position of steel in concrete	Unreliable when:
	– rebar closely spaced, different types/sizes, at deep cover
	– site-specific calibrations not done
	– other magnetic material nearby (windows, bolts, conduits)
	(*Note*: austenitic stainless steels are non-magnetic)
Chloride testing: chemical analysis	Chlorides in aggregates give misleading results
	Chlorides in cracks or defects difficult to determine accurately
	Slag, concretes difficult to analyse
	Large samples required to allow for the presence of aggregates

Carbonation depth: chemical method (pH indicator)

- Slightly underestimates carbonation depth
- Difficult to discern colour change caused by pH indicator in dark-coloured concrete
- Indicator ineffective at very high pH levels (e.g. after electrochemical re-alkalisation)
- Testing must be done only on very freshly exposed concrete surfaces (before atmospheric carbonation occurs)

Rebar potentials: potentiometer (voltmeter) using copper/copper sulphate reference electrode

- Not recommended for carbonation-induced corrosion
- Interpretation is a specialist task
- Delamination could disrupt potential field, and thus produce false readings
- Environmental effects (temperature, humidity) influence potentials
- No direct correlation between rebar potential and corrosion rates
- Stray currents influence measured potentials

Resistivity: Wenner probes and resistivity meter

- Carbonation and wetting fronts affect measurements
- Concrete with high contact resistance at surface results in unstable readings
- Rebar directly below probe influences readings

Corrosion rate: linear polarisation resistance (galvanostatic linear polarisation resistance with guard-ring sensor)

- Sophisticated technique, requires considerable expertise to operate
- Environmental and material conditions have large influence on measurements and single readings are generally unreliable

Table 1.5 *Visual assessment of structural failure: items for checklist*

Item	Details
Background data	
Identification	Reference, number, location
Environment	Severity and type of exposure
History	Age, design data, repairs
Original condition	
Surface condition	Honeycombing, bleeding, voids, pop-outs
Early cracking	Plastic settlement or plastic shrinkage
Concrete quality	Surface hardness, density, voids, colour
Rebar cover	Covermeter survey, mechanical breakout
Structural effects	Overloading, dynamic effects, structural cracking
Present condition	
Surface damage	Abrasion, staining, chemical attack, spalling, leaching
Staining	Rebar corrosion, alkali–aggregate reaction gel, efflorescence, salts
Cracking	Width, pattern, location, causes of cracking
Joint deficiencies	Joint spalls, vertical and lateral movements, seal damage
Rebar condition	Visual examination of bar, rust and pitting damage
Carbonation	Indicator test on cores or mechanical breakouts
Delamination	Size, frequency, severity of delamination
Previous repairs	Integrity of repairs, signs of damage near repair locations

commonly determined as acid soluble or total chlorides in accordance with the appropriate national standard.

Chloride sampling and the determination of the chloride level in concrete are illustrated in Figure 1.2 and are usually done in the following manner:

1. Concrete samples are extracted as either core or drilled powder samples.
2. Depth increments are chosen depending on the cover to steel and the likely level of chloride contamination (increments are typically between 5 and 25 mm).
3. Dry powder samples are digested in diluted nitric acid to release all chlorides.

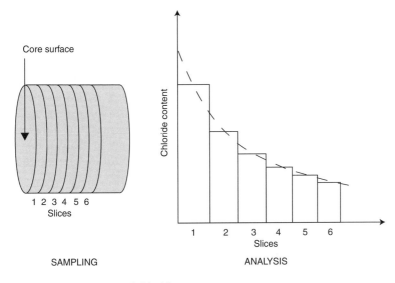

Figure 1.2 *Determination of chloride content*

4. Chlorides are analysed using potentiometric titration or the Volard method.
5. Chloride contents should preferably be expressed as a percentage by mass of cement.
6. Chloride profiles may be drawn such that chloride concentrations may be interpolated or extrapolated for any depth.
7. Future chloride levels can be estimated from Fick's second law of diffusion.

The corrosion threshold depends on several factors, including concrete quality, cover depth and saturation level of the concrete. The probability of corrosion may be assessed from the qualitative rating shown in Table 1.6 for acid-soluble chloride contents.

Table 1.6 *Qualitative risk of corrosion based on chloride levels*

Chloride content by mass of cement (mass %)	Probability of corrosion
<0.4	Low
0.4–1.0	Moderate
>1.0	High

Carbonation depth
Carbonation depth is measured by spraying a fresh fracture surface of the concrete with a phenolphthalein indicator solution (1% by mass in ethanol/water solution). Phenolphthalein remains clear where concrete is carbonated but turns pink/purple where concrete is still strongly alkaline (pH > 9.0). Carbonation moves through concrete as a distinct front and reduces the natural alkalinity of concrete from a pH in excess of 12.5 to approximately 8.3. Steel starts to depassivate when the alkalinity is reduced below pH 10.5. The progress of the carbonation front is shown in Figure 1.3. For prediction purposes, the rate of carbonation is approximately proportional to the square root of time.

Environmental conditions most favourable for carbonation (i.e. 50–65% relative humidity (RH)) are usually too dry to allow rapid steel corrosion, which normally requires humidity levels above 80% RH. Structures exposed to fluctuations in moisture conditions of the cover concrete, such as may occur during rainy spells, are however vulnerable to carbonation-induced corrosion.

Rebar potentials
Chloride-induced corrosion of steel is associated with anodic and cathodic areas along the rebar with consequent changes in the electropotential of the steel. It is possible to measure these rebar potentials at different points and plot the results in the form of a *potential map*. Measurement of rebar potentials may determine the thermodynamic risk of corrosion, but cannot evaluate the kinetics of the reaction. Rebar potentials are normally determined in accordance with ASTM C876: 1991 using a copper/copper

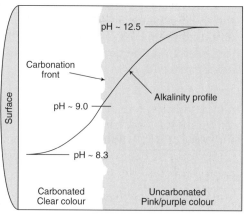

Sliced core

Figure 1.3 *The progress of the carbonation front*

Table 1.7 *Qualitative risk of chloride-induced corrosion*

Rebar potential (-mV Cu/CuSO4)	Qualitative risk of corrosion
<200	Low
200–350	Uncertain
>350	High

sulphate reference electrode connected to a hand-held voltmeter. The qualitative risk of corrosion based on rebar potentials is given in Table 1.7. Note that the technique is not recommended for carbonation-induced corrosion where clearly defined anodic regions are absent.

The procedure for undertaking a rebar potential survey is as follows:

1. Mark up a grid pattern in the area of measurement (not more than 500 mm centres).
2. Make an electrical connection to clean steel by coring or breaking out concrete.
3. Use a multimeter to check that the steel is electrically continuous over the survey area.
4. Wet the concrete surface with tap water if the concrete appears to be dry.
5. Take and record readings either manually or by using a data logger.
6. Check the data on site to ensure these correlate with the visual signs of corrosion.

Rebar potential measurements are relatively quick to perform. Absolute values are often of lesser importance than are the differences between values measured on a structure. A shift of several hundred millivolts over a short distance of 300–500 mm often indicates a high risk of corrosion.

Resistivity
Concrete resistivity controls the rate at which steel corrodes in concrete once favourable conditions for corrosion exist. Resistivity is dependent on the moisture condition of the concrete, on the permeability and interconnectivity of the pore structure, and on the concentration of ionic species in the pore water of concrete.

- poor quality, saturated concrete has low resistivity (e.g. <10 kΩ-cm)
- high-quality, dry concrete has high resistivity (e.g. >25 kΩ-cm).

Resistivity measurements are simple to perform on site and are done with a Wenner probe connected to a portable resistivity meter. The outer two probes send an alternating current through the concrete, while the inner two probes measure the potential difference in the concrete. Once the concrete resistivity is known, a rough assessment of likely corrosion rates can be made as shown in Table 1.8. This assessment assumes that conditions are favourable for corrosion.

Corrosion rate measurements
Corrosion rate measurements are the only reliable method of measuring actual corrosion activity in reinforced concrete. A number of sophisticated corrosion monitoring systems are available, based primarily on linear polarisation resistance (LPR) principles. Corrosion rate measurements on field structures are most commonly done using galvanostatic LPR techniques with a guard-ring type sensor to confine the area of steel under test. Table 1.9 shows a qualitative guide for the assessment of corrosion rates of site structures.

Repair strategies

Repair of reinforced concrete structures needs to be undertaken in a rational manner to guarantee success. An increasing number of repair options are available that must be considered in terms of cost, technical feasibility and reliability. Engineers need to understand all the relevant material, structural and environmental issues associated with concrete repairs in order to make intelligent choices.

Table 1.8   *Likely corrosion rate based on concrete resistivity*

Resistivity (kΩ-cm)	Likely corrosion rate for given corrosive conditions
<12	High
12–20	Moderate
>20	Low

Table 1.9   *Qualitative assessment of corrosion rate measurements*

Corrosion rate (μA/cm^2)	Qualitative assessment of corrosion rate
>10	High
1.0–10	Moderate
0.2–1.0	Low
<0.2	Passive

Various factors determine the suitability and cost-effectiveness of repairs:

- level of deterioration
- specific conditions of the structure
- environmental conditions.

Therefore, high-quality repairs require a thorough investigation of the causes of deterioration, appropriate repair specifications and competent execution of the repair work. This can only be done when independent experts carry out structural investigations, engineers with specialist repair expertise draw up specifications and competent contractors undertake repairs. The various repair options are compared in Table 1.10.

Patch repairs
The approach to repairing damaged concrete structures depends on whether the corrosion is carbonation induced or chloride induced. The two types of corrosion are contrasted in Table 1.11.

Important aspects of patch repair procedures are to:

- fully expose all corroded reinforcement by removing all cracked and delaminated concrete
- thoroughly clean the corroded reinforcement and apply a protective coating to the steel surface (e.g. anti-corrosion epoxy coating or zinc-rich primer coat)
- coat or seal the entire concrete surface to reduce moisture levels in the concrete.

Patch repair has limited success against chloride-induced corrosion as the surrounding concrete may be contaminated with chloride and the reinforcement is therefore still susceptible to corrosion. The patched area of new repair material often causes the formation of incipient anodes adjacent to the repairs, as shown in Figure 1.4.

These new corrosion sites not only affect the structure but often also undermine the repair, leading to accelerated patch failures in as little as 2 years. Consequently, it is necessary to remove all chloride-contaminated concrete in the vicinity of the reinforcement.

Complete removal of chloride-contaminated concrete should successfully halt corrosion by restoring passivating conditions to the reinforcement. Mechanical removal of cover concrete is usually done with a pneumatic hammer, hydro jetting or abrading machines. This form of repair is most successful when treating areas of localised low cover, before significant chloride penetration has occurred. If repairs are only considered once corrosion damage has become fairly widespread, it will be expensive to remove mechanically the chloride-contaminated concrete from depths well beyond the reinforcement.

Table 1.10 *Repair strategies*

Strategy	Comments
Patching: removal of all cracked and delaminated concrete and cleaning of all corroded reinforcement, application of protective coatings to steel and repairing with mortar or micro-concrete	Popular due to low cost and temporary aesthetic improvement. Limited success against chloride-induced corrosion
Barrier coatings: these systems attempt to seal the surface of the concrete, restricting the flow of oxygen to the cathode, thus stifling corrosion	Not practical for large concrete structures, because large amounts of oxygen are already present in the system. Generally ineffective due to the presence of defects in the new coating or damage in service. Likely to promote the formation of differential aeration cells, further exacerbating the corrosion potential
Hydrophobic coating (penetrating pore liners, e.g. silane and siloxane): surface capillary channels are lined with a hydrophobic coating, which repels water during wetting but allows water vapour movement for drying	Reduces the moisture content, and thereby electrolytically stifles the corrosion reaction. Suitability for marine structures is questionable due to the high ambient humidity, capillary suction effects and the presence of high salt concentrations, all of which interfere with drying. Applied to a new construction is effective for about 10–15 years
Migrating corrosion inhibitors: organic-based materials (e.g. amino-alcohol) suppress corrosion by being adsorbed onto the steel surface and displacing corrosive ions, such as chlorides, interfering with the anodic dissolution of iron and simultaneously disrupting the reduction of oxygen at the cathode	Effectiveness of inhibitors controlled by environmental, material and structural factors. Migrating inhibitors penetrate by vapour diffusion. Movement is fairly rapid through partially saturated concrete, but penetration is poor in near-saturated concretes (e.g. partially submerged marine structures with high moisture and salt levels)

Control of chloride-induced corrosion is largely dependent on the chloride levels at the reinforcement

Effectiveness of inhibitors is enhanced when they are used in combination with hydrophobic coatings

Electrochemical techniques: restore the passivated condition of the steel by the temporary application of a strong electric field to the cover concrete region

Re-alkalisation: restoring the alkalinity of carbonated concrete non-destructively; treatment can be completed in less than 2 weeks

Electrochemical chloride removal (ECR): a more time consuming and complex technique; its suitability must be carefully assessed

Cathodic protection: the electrical potential of the embedded reinforcement is artificially increased either by an impressed external current or by a sacrificial anode system, thus decreasing the corrosion rate of the steel

Sacrificial anode system: most effective in submerged structures (concrete saturated and resistivity low) and temperatures above 20°C

Impressed current: anode systems designed for long life (20–50 years)

Cathodic protection systems require electrically continuous reinforcement and uniformly conductive, delamination-free concrete cover

Demolition/reconstruction: only viable if deterioration of the total structure is very advanced

Corrosion damage is generally confined to the near-surface regions of a structure and surveyors must guard against overestimation of damage

Table 1.11 *Comparison of carbonation-induced and chloride-induced corrosion*

Carbonation-induced corrosion	Chloride-induced corrosion
General corrosion with multiple pitting along the reinforcement	Localised severe pitting corrosion with distinct anode and cathode sites
Carbonated concrete tends to have fairly high resistivity, which discourages macro-cell formation, and thus corrosion rates are moderate	High salt concentrations in the cover concrete mean that macro-cell corrosion is possible, with relatively large cathodic areas driving localised intense anodes, resulting in high corrosion rates
External signs of corrosion that can be easily identified visually (e.g. surface stains, cracking or spalling of concrete)	Much of the reinforcement may be exposed to corrosive conditions, but only localised anodic regions will show visible signs of distress with time
Repairs are generally successful, provided all the corroded reinforcement is treated	Chloride-induced corrosion is far more pernicious and difficult to treat than is carbonation-induced corrosion

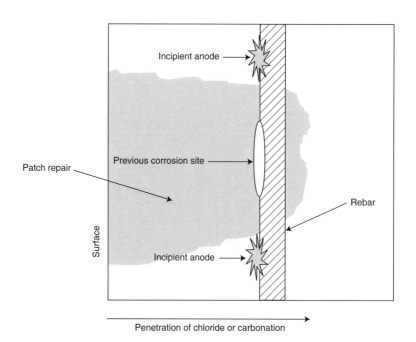

Figure 1.4 *Formation of incipient anodes after patch repairs*

Migrating corrosion inhibitors

A corrosion inhibitor is defined as a chemical substance that reduces the corrosion of metals without a reduction in the concentration of corrosive agents. The effectiveness of migrating corrosion inhibitors is generally controlled by environmental, material and structural factors (Table 1.12). It is critical that satisfactory penetration of corrosion inhibitors is checked before undertaking full-scale repairs. Effective inhibition may not be possible if the chloride levels by mass of cement are above 1.0% at the reinforcement. Better inhibition is possible if treatment is done earlier when chloride contents are lower.

Electrochemical techniques

Re-alkalisation. The electrochemical treatment consists of placing an anode system and sodium carbonate electrolyte on the concrete surface and applying a high current density (typically 1 A/m^2). The electrical field generates hydroxyl ions at the reinforcement and draws alkalis into the concrete.

Electrochemical chloride removal. Chloride removal is induced by applying a direct current between the reinforcement and an electrode that is placed temporarily onto the outside of the concrete. The impressed current creates an electric field in the concrete that causes negatively charged ions to migrate from the reinforcement to the external anode. The technique decreases the potential of the reinforcement, increases the hydroxyl ion concentration and decreases the chloride concentration around the steel, thereby restoring passivating conditions. Figure 1.5 shows the basic principles of electrochemical chloride removal (ECR).

Table 1.12 *Likely performance of migrating corrosion inhibitors in concrete*

Corrosive conditions	Concrete conditions	Severity of corrosion	Likely inhibition
Mildly corrosive, low chlorides or carbonation	Dense concrete with good cover depths (>50 mm)	Limited corrosion with minor pitting of steel	Good
Moderate levels of chloride at rebar (i.e. <1%)	Moderate quality concrete, some cracking	Moderate corrosion with some pitting	Moderate
High chloride levels at rebar (i.e. >1%)	Cracked, damaged concrete, low cover to rebar	Entrenched corrosion with deep pitting	Poor

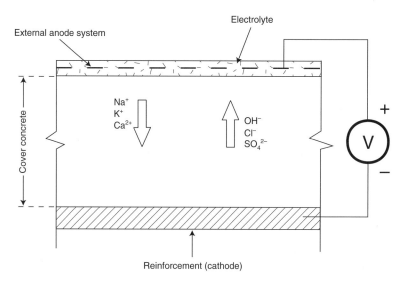

Figure 1.5 *The ECR technique*

The effectiveness of ECR depends on several factors:

- the extent of chloride contamination in the concrete
- the structural configuration, including the depth and spacing of the reinforcement
- the applied current density and the time of application
- the pore solution conductivity and the resistance of the cover concrete
- the presence of cracks, delamination and defects causing uneven chloride removal.

ECR typically takes 4–12 weeks to run at current densities within the normal range of 1–2 A/m². In some circumstances, chlorides beyond the reinforcement may be forced deeper into the concrete during the process. There is a risk that chlorides left in the concrete may diffuse back to the reinforcement and cause further corrosion with time.

The feasibility of using ECR depends on a number of factors:

- the presence of major cracking, delamination and defects that will require repair before ECR
- large variations in reinforcement cover that will cause differential chloride extraction and possible short-circuiting
- reactive aggregates require special precautions to avoid possible alkali–silica reaction (lithium salts should be used in these cases)
- prestressed concrete structures may be susceptible to hydrogen embrittlement after ECR (special precautions are needed to eliminate this risk)

- temporary power supplies of significant capacity are required during the application of ECR.

Cathodic protection systems

Cathodic protection (CP) systems have an excellent track record in the corrosion control of steel and reinforced concrete structures. In *sacrificial anode systems* the anode consists of metals higher than steel in the electrochemical series (e.g. zinc). The external anode corrodes preferentially to the steel and supplies electrons to the cathodic steel surface.

Impressed current CP systems use an external electrical power source to supply electrons from the anode to the cathode. The anode is placed near the surface and is connected to the reinforcement through a transformer rectifier that supplies the impressed current (Figure 1.6). Anodes may be conductive overlays, titanium mesh within a sprayed concrete overlay, discrete anodes or conductive paint systems.

CP repair of concrete structures requires a thorough corrosion survey by a specialist and the design needs to be undertaken by a corrosion expert. Reliable CP systems are fully controlled and monitored by a series of embedded sensors in order to ensure optimum performance. This is essential since under- or overprotection of the reinforcement may be potentially harmful to the structure or the CP system. Continuous monitoring of CP systems is usually done remotely by modem and the power consumption during operation is extremely small.

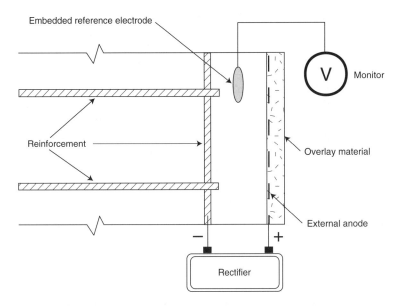

Figure 1.6   *A typical cathode protection system*

Demolition or reconstruction

This option should only be considered as a last resort since the total cost (capital costs plus loss of service and temporary works) is usually well in excess of repairs costs. Engineers who have limited repair experience or lack confidence in new repair systems often prefer demolition and reconstruction. It is crucial nevertheless that lessons are learnt from the old structure when designing the replacement.

Alkali–aggregate reaction

Alkaline pore solutions in the concrete react in moist conditions with certain types of aggregate to form an expansive gel, resulting in the internal disruption of the concrete. The reaction is slow and its effects only become noticeable after several years of service. Depending on the severity of the attack, the consequences of alkali–aggregate reaction (AAR) are:

• degradation of appearance
• deterioration in strength
• decrease in durability.

Types of alkali–aggregate reaction

Alkali–silica reaction

Alkali–silica reaction (ASR) is the most common form of AAR. Alkaline pore solutions react with metastable or highly disordered silica phases (opal, cristobalite, trydimite, volcanic glass, as well as highly strained, microcrystalline or cryptocrystalline quartz) found in particular silicate aggregates (quartzite, greywacke, argillite, hornfels, phyllite, granite, granite-gneiss, granodiorite, etc.) to form a reactive gel that expands when it imbibes water. When the swell pressure exceeds the tensile strength of the concrete (about 4 MPa) the concrete disrupts internally.

Alkali–carbonate rock reaction

Alkali–carbonate rock reaction (ACR) is seldom found. The alkaline pore solutions react with certain carbonate rocks (argillaceous dolomitic limestone), but no expansive gel is formed. The expansion is believed to be the result of dedolomitisation (reaction between alkali hydroxides and dolomite crystals) and additional swelling of clay minerals in the limestone made possible by the increased permeability of the rock.

Recognition of alkali–aggregate reaction

It should always be borne in mind that some indicators of AAR may be the result of other mechanisms, operating either in conjunction or on their own. Detailed analysis of the structure is thus imperative.

Cracking of concrete

The main and most obvious indicator of AAR is cracking of the concrete. The cracks become just noticeable after about 5 years, but may develop with time to fissures more than 1 mm wide. Where mass concrete is relatively free to expand in all directions, map or pattern cracking develops, but restraint in any direction will influence this crack pattern.

In reinforced concrete the cracking tends to follow the orientation of the main reinforcement and maximum stress directions. If the cracks extend to the reinforcement, the steel will start to corrode because the protective effect of the cover concrete has been nullified. Rust stains will start to appear from the cracks, suggesting that corrosion of the reinforcement is the cause of the crack formation. However, in this case the appearance of rust stains is the result, not the cause, of crack formation.

Expansion of concrete members

Expansion of the concrete results in the closure of movement joints. Warping and offsetting of structural members could also develop.

Presence of gel

In severe cases, streaks and drops of gel with a resinous, jelly-like appearance (sometimes stained) develop on vertical surfaces. Beware of confusion with carbonated lime efflorescence. If the area is treated with a solution of uranium acetate the ASR gel will be fluorescent under UV light. This is a specialist test and the results must be interpreted with caution.

Discoloration

Areas with a well-developed crack pattern may appear dark, giving the impression of permanent dampness. In severe cases, actual dampness could develop on the surface.

Confirmation of alkali–aggregate reaction

Suspected cases require full macroscopic and microscopic examination of core samples, not loose fragments taken from the surface. The cores must be at least 100 mm in diameter and 100 mm long. The exact position of sampling point, direction of drilling, location and condition of the reinforcement must be recorded. The crack pattern in the immediate vicinity of the coring location must be noted.

Various features are revealed by macroscopic investigation of cores sliced longitudinally:

- dark staining along cracks through aggregate, generally parallel with the surface of the concrete mass

- a white deposit on fracture surfaces of the aggregate, concentrated around the periphery and 0.3–1 mm thick, giving the impression of a reaction rim
- voids in the concrete and cracks in the aggregate or paste are lined or completely filled with a reaction product, which is translucent or porcelaineous, hard or soft.

Positive confirmation should be based on thin section petrographic microscopy, x-ray diffraction (XRD) analysis of the reaction product and scanning electron microscopy (SEM) investigation with energy-dispersive analysis.

Conditions necessary for alkali–aggregate reaction

High alkalinity of pore solution
The main contributor to the alkalinity of the pore solution is soluble alkali in the clinker component of the Portland cement. Minor sources are alkalis derived from reactions between the lime formed during hydration of Portland cement and alkali-containing minerals in the aggregate, as well as external sources such as salts in aggregates, mixing water, certain admixtures and seawater spray. The minimum alkali content in the concrete that is required for ASR to develop varies between about 2 and 4 kg/m^3, depending on the aggregate type.

Reactive phases in the aggregate
There is a large variation in the reactivity, not only between different types of aggregate materials, but also within a particular type of rock. The reactivity is determined from service records or laboratory tests for each source of aggregate.

Environmental conditions
The reaction rate roughly doubles with each 10°C increase in mean annual ambient temperature. ASR requires an internal relative humidity of more than about 75%. Climatic fluctuations increase the reaction rate.

Preventive measures

- Use only non-reactive aggregates (not always available).
- Use only low alkali cements (the only available cements may be high in alkalis).
- Use appropriate extenders (experimentation is needed to optimise effectiveness).
- Prevent continued wetting of the concrete (often unfeasible in practice).

Remedial action

- Ascertain whether expansion has stopped.
- Monitor growth in crack widths.
- Monitor expansion and deflection of structural elements.
- Determine the potential future expansion by accelerated expansion tests.

Reaction dormant
- Fill cracks with suitable grout or filler.
- Aesthetic considerations may prescribe coating of repaired surfaces.

Reaction active
Immediate crack repair will serve no purpose. Continued expansion must first be prevented by curtailing the reaction. Two approaches are available.

- Treat the affected parts of the structure with silane or siloxane to inhibit the ingress of water. Such treatment allows the concrete to dry out over time, thus stopping the reaction, provided that the ambient relative humidity is below about 75%. (**See also: Concrete, Repair strategies, p. 14**)
- Replace the alkali by lithium to eliminate the ASR. A lithium-based impregnation treatment is presently under trial.

Important

- If the durability of the concrete is affected to such a degree that there is concern for reinforcing steel corrosion, the structure should first be treated for corrosion, such as with a migrating corrosion inhibitor. (**See also: Concrete, Repair strategies, p. 14**)
- Where possible, the structure should be isolated from free water or wet soil by effective barrier systems. If appropriate, install ventilated cladding to protect the structure from rain.

Chemical attack on concrete

Concrete is vulnerable to attack by naturally occurring solutions or various industrial wastes (effluent, sugars, lactic acid, etc.). All liquids with a pH below 12.5 will attack concrete, but the attack will be slow if the pH is above about 6, and increases rapidly with increasing acidic conditions. The rate of reaction depends on the temperature, flow rates, solubility of the reaction products, mobility of the ions and permeability of the concrete. Disruption of the concrete is classified as either leaching or spalling damage, caused by surface or internal deterioration mechanisms.

Soft water

Attack mechanism

Softwater attack is mainly a problem associated with large civil engineering structures, such as irrigation schemes. However, water supply conduits made from fibre–cement composites may fail as a result of softwater attack.

Softwater attack can be seen in watercourses where the aggregate of canal concrete is slowly exposed in the water flow region. When the reaction extends through the canal lining, major leakage paths form that may wash away backfill material and cause the collapse of the structure.

> Soft water is defined as water deficient in dissolved calcium and magnesium ions. Some soft water (also termed 'pure water') contains aggressive carbon dioxide or organic acids such as humic acids in solution, reducing the pH to below 5.0 and thus making the water highly aggressive.

Remedial action

- There are no remedial actions to halt or reverse the action of soft water.
- Pure or slightly acidic domestic water supplies should be stabilised to avoid the cement being dissolved.
- It is not possible to make acid-proof cement products using Portland cement. Attempts to buffer the action by using dolomite as an aggregate will not be effective, because the aggressive waters will attack all alkaline material. The reactions will continue until all the cement paste has been dissolved.

Sulphates

Attack mechanism

A major cause of deterioration of concrete is when sulphate solutions react with certain constituents in set concrete and the process is accompanied by a volume increase. Reactions of this kind are when concrete is exposed to sulphates. It is important to note that these reactions require high moisture levels.

Three reactions involving sulphates may take place:

- Reaction of sodium sulphate with free calcium hydroxide in set concrete to form calcium sulphate (gypsum). The crystallisation of the reaction product results in expansion and disruption of the concrete.
- Reaction of calcium and sodium sulphates with hydrated calcium aluminates and ferrites to form calcium sulpho-aluminate (ettringite) and sulphoferrite hydrates. The reaction results in a doubling of the volume of the components, causing disruption of the concrete.
- The presence of magnesium sulphates results in the decomposition of hydrated calcium silicates, which results in a decrease in the strength of the concrete.

The sources of sulphates are:

- sulphate minerals (mainly gypsum) in the surrounding soil
- brackish groundwater and surface water
- sulphate-bearing solutions, such as domestic and industrial effluent.

Remedial action
- There are no remedial actions to reverse the disintegration of concrete caused by the disruptive action of sulphate attack.
- Sulphate attack will be curtailed if the continuous wetting of the concrete is prevented.
- Resistance of concrete products to attack by sulphates may be improved at the manufacturing stage by:
  - Autoclaving (6 hours at 850 kPa, 175°C). This is only practical in the case of certain precast units.
  - Using sulphate-resistant Portland cement. Rapid cooling of clinker minimises the formation of C_3A precipitation.
  - Incorporating fly ash and ground granulated blast furnace slag (GGBS) of appropriate quality and appropriate level into the mix.

See also:
Masonry, Sulphate attack, p. 59

Fire damage

Effect of high temperatures

High temperatures break down hydraulic cementitious compounds. The strength of concrete is not severely affected by temperatures below 250°C, but drops to 80% at about 450°C and to 50% at about 650°C. On cooling there is a further loss in strength. Note that fire temperatures in enclosed spaces can exceed 1500°C.

Concrete exposed to fires is prone to spalling because the low thermal conductivity results in steep temperature gradients. High moisture contents in the concrete also cause high pore pressures from trapped steam that exacerbate the risk of spalling. Surface damage starts with slight crazing and progresses to widespread scaling, cracking and extensive spalling. However, because of the low thermal conductivity of concrete, damage is usually confined to the outer 50 mm or so, leaving the underlying concrete in a sound condition, largely unaffected by the fire.

Fire-affected concrete has a lumpy, powdery appearance. The colour of fire-exposed concrete provides some indication of the maximum exposure temperature (Table 1.13).

Remedial action

- Remove all fire-damaged material; treat any exposed reinforcement and patch with repair mortar or micro-concrete.
- It is very important to realise that the repaired surface is not load bearing. It is thus vital to establish whether the structural members still have adequate load-bearing capacity after removing all fire-damaged material.

See also:
Concrete, Repair strategies, p. 14

Frost attack

Damage mechanism

Dry concrete (i.e. concrete with a low level of water in the interconnected pores) is resistant to frost action. The gel pores in hydrated Portland cement are so small that the water in them will not freeze. The freezing point of the water in the capillaries depends on the salt content, but it is well below 0°C.

When the water in the interconnected pores of wet concrete freezes, there is a 12% increase in volume. If these pores are saturated above a critical value (about 80–90%, depending on the properties of the concrete) frost action could disrupt the concrete. The damage caused by cycles of

Table 1.13 *Colour of concrete exposed to high temperatures*

Temperature (°C)	250	300	600	950
Colour	Pink	Pink-red	Black-grey	Buff

freezing and thawing is cumulative and leads to flaking of the surface layers or even total crumbling of the material.

The mechanisms causing disintegration of porous materials by frost action are quite complex. Pore structure and size distribution have a very decisive influence on the process.

Remedial action

• Unless the concrete is of poor quality and very porous, frost damage is unsightly, rather than damaging.
• In conditions where cycles of freezing and thawing are common, quality concrete (low permeability) should be used and compacted well during placing.
• Saturation of the concrete during thawing periods should be prevented.
• Air-entrained concrete could have good frost resistance, because the small voids formed by the air bubbles act as a pressure-relieving system.

See also:
Masonry, Frost attack, p. 65
Tiling, Frost attack, p. 78

Dimensional change

Stress failure (cracking or crushing) of concrete is not caused by material failure, but is the result of bad design or poor workmanship. The effects of shrinkage, creep and thermal expansion in large reinforced concrete structures must be allowed for in the design stage by incorporating movement joints, while good workmanship must ensure that these joints are effective.

Differential dimensional change in composite building systems (concrete frame structure with infilling of fired clay brickwork, or ceramic tiled areas on concrete substrates) contributes to the failure of such systems. Ceramic elements (clay bricks and tiles) tend to expand irreversibly with time, as a result of reaction with moisture. The combination of shrinkage and creep of concrete and expansion of ceramic elements can cause considerable stresses in a structure. The magnitude of such dimensional changes must be taken into account in the design, position and maintenance of movement joints.

Note that bleeding and the initial plastic shrinkage during the setting stage are not considered in this context.

Shrinkage and creep

Drying shrinkage

Apart from the initial plastic shrinkage that occurs during the setting process, cement-based materials shrink further when water is lost from the capillaries and gel pores in the hardened cement paste. The magnitude of drying shrinkage depends on the aggregate type and on the proportion, mix design and time. Drying is a slow process and depends on the size of the concrete unit and climatic conditions.

Very generally, concrete will shrink about 0.02% linearly (or 0.2 mm/m) during the first 4 weeks after it has been poured. As drying continues, the concrete could shrink linearly by another 0.02% over the following year or two. It should be remembered that the reversible expansion and shrinkage due to the wetting and drying of concrete (e.g. caused by rain) will be superimposed on the drying shrinkage movement.

Carbonation shrinkage

Reaction between atmospheric carbon dioxide and the hardened cement paste results in a reduction in volume of the paste. The shrinkage is a function of relative humidity, being highest at intermediate humidities. If the concrete is very wet, carbon dioxide cannot penetrate into the concrete, while the absence of water in very dry concrete restricts the reaction.

The development of this irreversible shrinkage is slow and in certain cases may exceed drying shrinkage in magnitude. Shrinkage cracks develop from the surface, decreasing the effective cover and leading to early reinforcement corrosion.

Creep

Cement-based materials creep under load. The dimensional changes of concrete during the short, medium and long term are very complex. Nevertheless, the combined effect of both drying shrinkage and creep could result in the long-term linear 'shrinkage' of concrete members by as much as 0.2% (or 2 mm/m).

Thermal expansion

Concrete and steel have very similar coefficients of thermal expansion. Therefore, differential expansion stresses in reinforced concrete are small

Table 1.14 *Thermal expansion of building materials*

Material	Coefficient of thermal expansion (per °C × 10⁶)	Expansion (mm/m per 20°C)
Steel	11–12	0.22–0.24
Aluminium	22–24	0.44–0.48
Dense concrete	11–13	0.22–0.26
Fired clay bricks and tiles	6–9	0.12–0.18
Calcium silicate bricks	8–14	0.16–0.28

and thus cannot cause stress failure. This is not the case with embedded aluminium fittings.

The thermal expansion of ceramic bricks and tiles is significantly higher than that of cement products (Table 1.14). However, the diurnal and seasonal temperature changes in structures are generally not sufficient to result in excessive thermal stressing of the structure.

Remedial action

- Incorporate and maintain effective movement joints in the structure.
- For repair of concrete that has suffered stress failure, see **Concrete, Repair strategies, p. 14**.

See also:
Concrete, Repair strategies, p. 14
Masonry, Differential movement, p. 53
Tiling, Differential movement, pp. 82, 84

Further reading

Australian Concrete Repair Association. *Guide to concrete repair and protection*. Standards Australia, Sydney, 1996.

Addis B.J. and Basson J.J. *Diagnosing and repairing the surface of reinforced concrete damaged by corrosion of reinforcement*. Portland Cement Institute, Midrand, 1989.

Addis B.J. and Owens G. (eds). *Fulton's concrete technology*, 8th edn. Cement & Concrete Institute, Midrand, 2001.

American Concrete Institute. *Guide to durable concrete*. American Concrete Institute, Detroit, 1992.

American Society for Testing and Materials. ASTM C876 Standard test method for half-cell potentials of uncoated reinforcing steel in concrete. ASTM, Philadelphia, 1991.

Broomfield J.P. *Corrosion of steel in concrete: appraisal and repair.* Chapman and Hall, London, 1997.

Comite Euro-International Du Beton. *Durable concrete structures.* Thomas Telford, London, 1992, Information Bulletin No. 183.

Concrete Society. *Diagnosis of deterioration in concrete structures.* Concrete Society, Crowthorne, 2000, Technical Report 54.

Hobbs D.W. *Alkali–silica reaction in concrete.* Thomas Telford, London, 1988.

National standard specifications and code of practices

Entry web sites

International Concrete Repair Institute: http://www.icri.org
Portland Cement Association: http://www.portcement.org

Chapter 2

Metal failure

Contents

Modes of metal failure **34**
Mechanical failure 34
Corrosion 34
    Tarnishing 35
    Atmospheric corrosion 35
    Galvanic (bimetallic) corrosion 35
    Crevice corrosion 36
    Pitting corrosion 36
    Stress corrosion 36
    Intergranular corrosion 37
    Erosion corrosion 37
Suppression of metal corrosion **37**
Passivation 37
Protection of exposed metal surfaces 37
Suppression of galvanic corrosion 38
Crevice and stress corrosion 38
Structural steel **38**
Atmospheric corrosion 39
Corrosion of steel embedded in concrete 39
Restitution of corroded steel components 41
Galvanised steel **42**
Atmospheric corrosion 42
Galvanic corrosion 42
Corrosion of galvanised reinforcing steel 43
Weathering steel **44**
Stainless steel **44**
Cleaning stainless steel 45
Repassivation 45
Painting of stainless steel 46
Aluminium **46**
Atmospheric corrosion 47
Galvanic corrosion 47
Corrosion at the concrete interface 47

Crevice corrosion 48
Anodised aluminium 48
Copper alloys **49**
Atmospheric corrosion 50
Galvanic corrosion 50
Dezincification of brass 50
Further reading **50**

Modes of metal failure

Metallic components in buildings invariably fail by one of two mechanisms, mechanical failure or corrosion, the latter being by far the more common mode.

Mechanical failure

Metals (in fact all materials) deform when they are stressed. If the applied stress does not exceed the yield stress of the metal, the metal will return to its original form when the stress is removed. However, if the yield stress is exceeded, then the metal will be permanently deformed. If the applied stress is continuously increased then fracture will occur when the ultimate tensile strength of the metal is exceeded. Examples of metal failure through excessive deformation or fracture are brass fittings permanently bent or metal cladding torn away from its fasteners. Metals may also fail when repeatedly stressed, for example the fatigue failure of door latches. Mechanical failure of metallic building products is not a common occurrence, because the working stresses can be established and allowed for by sound design.

Corrosion

Corrosion may be defined as the deterioration of a material due to the effect of the environment. Corrosion is generally perceived to be associated with moisture, but in reality the corrosive environment can be either dry or wet. The fundamental cause of corrosion is the inherent instability of metals in the metallic form. The tendency is for metals (except the noble metals) to revert to more stable forms, such as oxides or sulphides. The driving force for this reversion to more stable forms differs from metal to metal and is expressed as the electrode potential of that metal under defined conditions.

Tarnishing

Tarnishing occurs when exposed metal is evenly attacked by a dry corrosive gas or liquid. As far as building materials are concerned, oxidation is the only tarnishing reaction of importance.

The reaction rate depends on the character of the oxide layer. The reaction rate will be linear if the reaction product (oxide scale) is porous or non-adherent. Oxygen has continuous and constant access to the metal surface and the reaction continues unabated.

The reaction slows down or stops with time, with progress being either parabolic or logarithmic, if the reaction product is non-porous and adheres fully to the metal surface.

Atmospheric corrosion

This is the most common form of corrosion. The process is electrochemical in nature and requires the presence of moisture. Attack is uniform over the total exposed surface, leaving a scale or deposit at or near the surface.

The corrosion rate is generally low and predictable, which means that adequate allowance can be made in the design to deal with the problem. A common example is the general rusting of steel structures when exposed to moist atmospheres.

Certain factors are likely to increase the rate of atmospheric corrosion:

* high relative humidity and high ambient temperatures
* accumulation of salts or pollutants on the metal surface.

Galvanic (bimetallic) corrosion

This type of corrosion occurs when two metals (or alloys with different compositions) are conductively in contact, while exposed to moisture or a corrosive solution. Under these circumstances, the more reactive (less noble) metal will corrode faster than it would if it was not connected to a more noble metal. Conversely, the more inert metal will be protected and thus corrode at a slower rate than it would on its own. The reactive metal in such a couple is called the anode and the inert metal the cathode of the galvanic corrosion cell.

Metals and alloys commonly used in construction can be ranked in a galvanic series according to their relative reactivity under the prevailing conditions.

For a marine environment the ranking for metals commonly used in buildings is as follows:

↑ **Less reactive (cathodic)**

Stainless steel (passivated)
Copper, bronze, brass
Tin, lead
Iron and steel
Aluminium
Zinc

↓ **More reactive (anodic)**

This ranking shows that steel screws will corrode (rust) in contact with brass fittings, because in that situation steel acts as the anode. However, steel screws will not corrode when used to fix aluminium fittings, because then it acts as the cathode and it is the aluminium that corrodes.

Crevice corrosion
Localised corrosion between two regions of the same metal piece is the consequence of a concentration difference in ions or dissolved gases in electrolyte, usually as a result of differential aeration. In these so-called *concentration cells* the metal in areas with lower oxygen concentration is anodic relative to the rest of the same metal piece.

Concentration cells are most commonly formed by localised stagnant conditions. Metal preferentially corrodes in shielded or occluded crevices (under bolt heads or washers, in threads of pipes or bolts, or under sediments and organic growth) and is then deposited as a hydroxide outside the crevice. In the case of iron, the deposit is initially soft rust, $Fe(OH)_2$, which slowly transforms to hard rust, $Fe(OH)_3$.

Pitting corrosion
This type of corrosion is insidious and particularly dangerous, as it frequently goes undetected until failure occurs. Pitting corrosion can be a major problem with reinforcing steel.

Pitting corrosion is most usually caused by the presence of chloride ions. Under suitable conditions the ions attack localised weak spots in the passive surface layer, forming micro-anodes in a large cathode area. Corrosion is thus concentrated in very small anode areas. Once a small pit has been formed, corrosion continues as for crevice corrosion, because of the stagnant conditions developing inside the pit.

Stress corrosion
Regions in the same piece of metal that are at higher stress levels than the surrounding metal act as anodes in corrosive conditions. For example, stress cells are formed in those regions where sheet steel has been folded,

or at cold-worked steel nail heads and points. Stress cells are also formed in fabricated steel structures at those contact points where bolts or rivets have been used to clamp plates or beams together.

Intergranular corrosion
Failure at welded joints may occur in certain metals as a result of intergranular (micro-scale) corrosion.

Erosion corrosion
Corrosion will be accelerated in erosive situations, because the passive layer is continuously removed through abrasion, thus exposing the unprotected active metal surface to corrosive attack.

Suppression of metal corrosion

Passivation

Passivation is a very important phenomenon and control mechanism in corrosion. Metals (with the exception of the noble metals, such as gold and platinum) tend to react with free oxygen to form various oxides. If the crystal structure of the particular oxide is fully compatible with the crystal structure of the metal, the oxide will bond at the atomic level with the metal to form a continuous and strongly adherent structure, termed a *coherent layer*. On the other hand, if the crystal structure is not compatible with the metal crystal structure, the oxide cannot form strong atomic bonds with the parent metal and will tend to detach and peel off. Such oxide layers are termed *incoherent layers*.

The formation of a coherent oxide film on exposed surfaces of a metal is termed *passivation*. Coherent oxide layers are highly protective, because they impede oxidation reactions by protecting the metal from contact with free oxygen. The layers are generally very thin and invisible (1–10 nm thick), insoluble, non-porous and self-healing under oxidising conditions. Metals become chemically active under reducing conditions, and thus passivity applies only to certain environmental conditions. Examples of metals that self-passivate very strongly are titanium, aluminium, chromium and chrome-containing stainless steels.

Protection of exposed metal surfaces

Suitable coatings may protect metal surfaces exposed to corrosive conditions. Some coating systems are very durable, while others need to be maintained regularly. Systems specific to certain metals are discussed below, while painting systems in general are dealt with in Chapter 6.

Suppression of galvanic corrosion

There are several factors that influence the rate of galvanic corrosion. The absence of any one factor is generally sufficient to eliminate, or at least substantially suppress, galvanic corrosion. Remedial actions are listed in Table 2.1.

Crevice and stress corrosion

It is not always possible to avoid the existence of crevices or stress regions. Corrosion control thus depends largely on the prevention of an electrolyte coming into contact with the potentially corrosive regions. This prevention is usually achieved by means of coating systems.

Structural steel

Structural steels are low-alloy steels specially developed to be easily welded and with good formability, strength and fair corrosion resistance. They are produced as hot- or cold-rolled sections for structural steelwork, or rods and wire for reinforced concrete. Corrosion is by far the main reason for failure of structural steel.

Table 2.1 *Suppression of galvanic corrosion*

Corrosion factor	Remedial action
Electron flow from anodic metal to cathodic metal	Electrically isolate dissimilar metals from each other. Use a non-conducting isolation tape on mating surfaces, and insulating sleeves or bushes around bolts and screws
Ionic flow from cathodic metal to anodic metal	Prevent the electrolyte (salt or acidic solution) from making contact with the anode, particularly near contacts with the cathodic metal
Ratio of anode to cathode surface areas	Decrease the corrosion rate (mass loss per unit area) by increasing the relative surface areas. Use large anode surfaces or decrease the area of cathodic metal (e.g. by painting or coating)
Corrosion potential of metal couple	When possible, choose metals closer to each other in the galvanic series for specific exposure conditions

The Fe^{2+} ions formed at the anode precipitate as hydroxide in the vicinity of the cathode. Further oxidation occurs to form rust, which is capable of expansion through the uptake of water:

$Fe^{2+} + 2(OH)^- \rightarrow Fe(OH)_2$ (ferrous hydroxide)

$4Fe(OH)_2 + O_2 + 2H_2O \rightarrow 4Fe(OH)_3$ (ferric hydroxide)

$2Fe(OH)_3 \rightarrow Fe_2O_3 \cdot H_2O + 2H_2O$ (hydrated ferric oxide)

The hydrated oxides may swell up to ten times the original volume of the steel. The type of corrosion product formed at the steel surface depends on the environmental conditions:

* Red or brown rust forms under high oxygen concentrations. It is flaky, relatively soft and easy to dislodge from the steel surface.
* Black rust forms under low oxygen concentrations. It forms a relatively dense and hard layer that may be difficult to remove from the steel surface.

Atmospheric corrosion

The rust layer formed on steel by atmospheric corrosion is not fully coherent. The corrosion rate of new surfaces is initially relatively rapid, but after a year or two it slows down to a fairly slow rate. In rural environments the long-term corrosion rate is about $5\,\mu$m/year, while in heavy industrial areas it may be as high as $100\,\mu$m/year.

The first signs of active corrosion are rust stains. As corrosion develops, blistering and foliation develops. Severe corrosion of steel window frames could result also in cracking of the glazing or buckling of the frame.

Corrosion of steel embedded in concrete

Steel is embedded in concrete or mortar as reinforcing steel (rebar), wall ties, fasteners, steel window frames, etc.

The passive ferric oxide film on embedded steel may be disrupted by a reduction in the alkalinity of the concrete (principally by carbonation) or

Steel will corrode freely when exposed to moisture and oxygen under ambient conditions. When it is embedded in concrete, however, the high alkalinity (pH 12.5 or higher) stifles corrosion by the formation of a passive ferric oxide film on the surface. The ferric oxide layer forms a dense, impermeable film that suppresses further corrosion by limiting the movement of cations and anions near the surface.

by the presence of aggressive ions such as chlorides and sulphates. Depassivation of the steel occurs as follows:

- in carbonated concrete, insufficient hydroxyl ions are available to repair damage in the passive film
- in salt contaminated concrete, chloride ions break down the passive layer at localised areas and shift the corrosion passivity barrier to pH values above the pH range of concrete, thus encouraging metallic dissolution at these points.

Once depassivating conditions exist in concrete, either by a reduction in alkalinity (pH < 10.5) or by the presence of sufficient chloride ions (the *corrosion threshold value*), corrosion may occur.

It is important to note that the establishment of depassivating conditions at the steel surface is the fundamental requirement for corrosion to start, while other factors (e.g. oxygen and moisture availability) will determine the rate of corrosion. The corrosion process of steel in concrete is shown schematically in Figure 2.1. The nature of steel corrosion in concrete depends on the local conditions at the surface of the bar. The four states of corrosion that may be defined are summarised in Table 2.2.

High resistivity concrete with relatively deep covers tends to favour *micro-cell corrosion*, where the anodes and cathodes are close together, causing localised pitting. In conductive concrete, contaminated with salts, more widely spaced anode and cathode sites are formed, causing *macro-cell corrosion*.

Only pitting and general corrosion present a threat to the reinforcement and the severity will depend on a number of internal and external

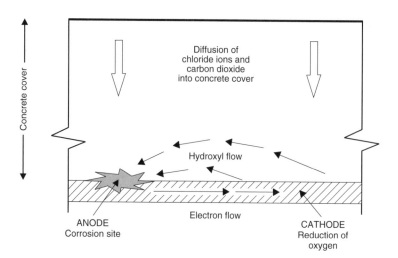

Figure 2.1   *The corrosion of steel embedded in concrete*

Table 2.2 *Types of corrosion of steel embedded in concrete*

Steel environment	Corrosion type
Steel embedded in sound, alkaline and uncontaminated concrete	*Passive corrosion state*: only minute levels of corrosion are needed to sustain the integrity of the ferric oxide film
Steel embedded in chloride-contaminated concrete	*Intense pitting corrosion*: local breakdown of the passive film, due to the presence of chloride ions, creates anodic condition and rapid corrosion. Adjacent steel acts as the cathode
Steel embedded in carbonated concrete	*General corrosion*: multiple pits along the steel surface result in an overall loss of passivity
Steel embedded in concrete that is mostly underwater	*Active, low potential corrosion*: insufficient oxygen is available to sustain the passive film, despite the high alkalinity of the concrete. The corrosion proceeds slowly

factors, which need to be assessed when conducting a corrosion survey. Internal factors include the concrete microstructure, cover depth and moisture condition. External influences, such as stray currents and microbial activity, may introduce a new dimension into the corrosion system, but are not considered here.

See also:
Concrete, Corrosion damage to reinforced concrete, p. 2

Restitution of corroded steel components

Structural steel (e.g. roof trusses, balustrades, window frames and pipes) is generally protected from atmospheric corrosion by suitable impervious coatings, such as paint (with or without inhibitors) and polymers (PVC powder coatings). If the integrity of the protective coating is compromised, differential aeration at the break will result in corrosion. Steel structures may also be protected by metal coatings (discussed below) and, rarely, by cathodic protection or sacrificial anodes.

During repairs to reinstate the integrity of the surface protection, it should be borne in mind that the spread of corrosion could be much wider than that indicated by surface signs:

*   The surface must be suitably prepared. All traces of corrosion products must be removed.

41

- If the breakdown of the surface protection is localised, sanding down those areas to bare metal could be sufficient. If the occurrence of corrosion is extensive, shot or grit blasting is recommended (wire brushing is ineffective to remove heavy rust).
- Round off sharp edges and scores, as they are not well covered by paint films.
- For surfaces to be painted, use a rust-inhibiting primer, followed by the recommended number of coats of quality paint. To avoid crevice corrosion, do not paint over crevices or lapping joints if there is still moisture trapped in them.
- For metal-sprayed surfaces, degrease and re-spray with the appropriate metal to the required thickness.

See also:
Paint, Metals, p. 109

Galvanised steel

The resistance of structural steel to general corrosion is considerably improved by galvanising. The thin layer of zinc is anodic relative to the steel and thus corrodes, protecting the steel. The methods of zinc coating steel are summarised in Table 2.3.

Atmospheric corrosion

Zinc has a reasonable durability, with corrosion rates ranging from about $0.5\,\mu$m/year in rural environments to more than $10\,\mu$m/year in heavily polluted industrial areas.

Soft water (containing dissolved oxygen and carbon dioxide) attacks zinc, while certain timbers (e.g. oak and red cedar) can corrode galvanised fittings.

Galvanic corrosion

Galvanised steel will corrode strongly if it comes in direct contact with copper alloys (tubing, fittings and plumber's debris) or copper-bearing solutions (rain dripping from copper power and telephone wires).

When zinc is protecting steel substrates as a sacrificial anode, the zinc is consumed more rapidly in the vicinity of exposed steel (e.g. at cut edges). In these areas the rate of zinc loss is much accelerated. The depletion of the zinc coating around the perimeters of galvanised roofing sheets and resultant early corrosion of the steel is a common occurrence.

Painting the galvanised steel extends protection markedly. Note that special primers are needed to prepare new zinc surfaces for painting.

Table 2.3 *Zinc coating methods*

Method	Description
Hot-dip galvanising	The component is dipped into molten zinc at 450°C. A zinc–iron alloy is formed at the interface, which ensures a good bond with the protecting zinc layer. The thickness of the zinc covering is generally 50–200 μm, but could be as little as 20 μm
Zinc spraying	The bond is mechanical, not chemical. Atomised molten zinc is sprayed onto a roughened surface, to a thickness of up to 200 μm. The process provides excellent protection and is suitable for on-site application to large structures
Electroplating	The process is restricted to small components or wire. The zinc layer has good flexibility
Sherardising	Small components (bolts and nuts) are tumbled in hot zinc dust, forming a thin layer of zinc with good adhesion. The degree of protection is determined by the layer thickness
Zinc-dust paint	These paints have a high loading of zinc particles (up to 90% zinc). They are generally used as prefabrication primers

See also:
Paint, Metals, p. 109

Corrosion of galvanised reinforcing steel

The use of galvanised reinforcing steel in marine structures must be treated with caution. Freshly poured concrete will attack the zinc coating, but the formation of calcium zincate (which is insoluble in alkaline solutions) soon inhibits this reaction. However, the hydrogen gas formed by the reaction not only reduces the bond strength between the rebar and the concrete, but also allows the easy penetration of seawater into the structure. The chlorides will rapidly deplete the zinc coating, resulting in serious corrosion damage to the concrete within a few years. The formation of hydrogen gas can be prevented by pretreating the galvanised rebars with chromate.

Weathering steel

Weathering steel, also called *corrosion-resistant steel*, has a superior corrosion resistance to carbon steel. Weathering steels contain between 0.25% and 0.55% alloying elements, mainly copper, chromium and phosphorus. On exposure, a dense, coherent, tenacious brown oxide film is formed on the surface that strongly inhibits the corrosion process. Weathering steel is popular for architectural sculptures or roofing of prestige buildings.

The atmospheric corrosion rate of weathering steel in unpolluted atmospheres is about a quarter of that of ordinary steel, but in heavily polluted conditions, particularly where there are high SO_2 levels, the corrosion rate will be similar to that of ordinary steel.

Stainless steel

Stainless steels are low-carbon alloy steels containing a minimum of 11% chromium. The chromium forms a tough, coherent and invisible chromium oxide film on the surface, imparting excellent corrosion-resisting properties to the steel. The crystal structure of the alloys is determined by their composition. The main groups of stainless steel alloys are the austenitic, ferritic, martensitic and duplex stainless steels.

There are considerable differences between the various stainless steels and each type has a specific field of application. The correct choice of type is thus vital to get the required durability and service from this material.

Only austenitic and (to a much lesser extent) ferritic stainless steels are used in building construction. Austenitic stainless steel is preferred for all external and internal architectural applications (e.g. street furniture, cladding, facades, shop fronts, signs), but ferritic stainless steel is suitable only for interior applications in non-aggressive environments. Martensitic and duplex stainless steels are not used as construction materials. The properties of austenitic and ferritic stainless steel are compared in Table 2.4.

The very strong and tenacious chromium oxide layer makes passivated stainless steel very corrosion resistant, even in heavily polluted industrial environments or seashore conditions. However, any iron-bearing particles in direct contact with stainless steel will rapidly rust by galvanic action. If the ferrous particles are very small (fine dust), a blush of rust may form that is difficult to identify. A rapid and reliable identification is given by the ferroxyl test:

- A solution of 3% potassium ferrocyanide and 3% nitric acid is wiped or sprayed onto the steel surface.
- Any ferrous contamination will show as a blue colour within 3–5 minutes.
- The test solution and colour are washed off with 3% acetic acid (vinegar).

Cleaning stainless steel

The cleaning methods appropriate for different conditions are listed in Table 2.5. The least aggressive procedure should always be used.

Repassivation

After aggressive cleaning or repair of damage it may be necessary to restore the integrity of the passive layer. Repassivation is done using pastes, gels or solutions containing *only* nitric acid (HNO_3), followed by thorough washing with clean water to ensure that all traces of acid are removed.

Table 2.4 *Comparison of austenitic and ferritic stainless steel*

Properties	Austenitic stainless steel	Ferritic stainless steel
Common types	304, 316 and 317	430 and 409
Basic composition	Chromium 18%	Chromium 12–18%
	Nickel 8%	Carbon content <0.01%
	Molybdenum 3%	
	Carbon content <0.10%	
Basic properties	Non-magnetic	Magnetic
	Weldability excellent	Weldability moderate to poor
	Corrosion resistance excellent	Corrosion resistance moderate to good
	Formability excellent	Formability fair to moderate
	Strength good, work hardens	Strength good, does not work harden

Table 2.5 *Cleaning procedures for stainless steel*

Condition	Properties
Routine cleaning	Mild detergent, soft cloth or sponge
Stubborn stains	Mild household abrasive cleaner, then routine clean
Grease, fats, oils	Water-soluble degreasing agent, and then routine clean
Rust marks	Swab with 10–15% dilution nitric acid, then routine clean

Painting of stainless steel

Hot-rolled mill finishes have limited aesthetic appeal, but they can be successfully painted. Surface preparation is critical, while twin-pack resin-based primers must be used to ensure full adhesion. Paint manufacturers should be consulted regarding the best generic paint system for the expected environmental conditions.

See also:
Paint, Metals, p. 109

Aluminium

Aluminium has good corrosion resistance in most atmospheres on account of the strong passive oxide layer that forms spontaneously on its surface and which is self-healing when damaged. Mill-finish sheet and extruded aluminium sections are used extensively in the building

---

Aluminium is seldom used in its pure form and is normally alloyed with small proportions of other elements, such as copper, manganese, silicon, magnesium and zinc. Alloying aluminium results in a wide range of materials with properties that differ greatly. The choice of the correct alloy for a specific application is vital to ensure durability.

Aluminium alloyed with magnesium and silicon is used for extruded sections. After controlled heat treatment (age hardening) these alloys have tensile strengths typically five times that of plain aluminium. The actual heat-treatment process is a clearly defined sequence of 'solution treatment', quenching and precipitation (or 'age') hardening. For building products these treatments are generally carried out at the mill.

---

industry for window frames, cladding, roofing, railing and exposed structural members.

Atmospheric corrosion

- The atmospheric corrosion rate for mill-finish aluminium is less than 10% of the rate for copper and less than 1% of the rate for structural steel.
- Mill-finish aluminium has a good resistance to industrial pollution, because the protective oxide layers are stable well into the acidic range.
- A small amount of pitting corrosion may appear on the surface if the SO_3 concentration is very high.
- The durability of aluminium is slightly reduced when exposed to salt-laden sea air.
- The extent of surface corrosion always appears to be worse than it actually is, because the volume of the corrosion product is many times greater than that of the metal consumed.

Galvanic corrosion

- Aluminium will corrode when in contact with structural steel, brass alloys and stainless steel.
- Aluminium in contact with galvanised steel will be protected because the zinc coating acts as the anode of the system. However, when the zinc has been depleted, the aluminium will become the anode and start to corrode.
- Aluminium should be isolated from dissimilar metals as far as possible.
- If metallic fasteners that are made of metals that are more noble must be used, the bimetallic corrosion of aluminium is generally low because of the large surface area of the aluminium section.
- Cathodic protection can be provided by zinc or magnesium anodes, or by an impressed potential.

Corrosion at the concrete interface

- Precautions should be taken where aluminium interfaces with concrete, whether embedded in fresh concrete or simply interfacing with set concrete. It is advisable to protect the aluminium with a protective coating (e.g. bitumen-based paint or alkaline wash).
- Some aluminium corrosion product at the interface with concrete may be tolerable provided that the structural integrity remains. As the corrosion product is stable and insoluble in water, it will often function as a barrier in its own right, and the corrosion rate will slow down.

- Splashes of alkaline building materials (mortar and concrete) cause surface damage in the form of spots or stains that are difficult to remove. Aluminium surfaces should be well protected during building operations.

Crevice corrosion

- Crevice corrosion is not a major problem with aluminium fabrications.
- The use of sealing compounds or adhesives, in conjunction with or as a substitute for rivets and screws, could assist in preventing crevice corrosion where sheet materials are joined to sections.
- 'Water spots' are formed when moisture penetrates between aluminium surfaces that are in contact during storage. The blemishes formed by this type of crevice corrosion are difficult to remove.

Anodised aluminium

Notwithstanding its excellent corrosion resistance under most environmental conditions, the surface of mill-finish aluminium is readily scratched, oxidises unevenly and develops pitting corrosion. Surface coatings, the most common and durable being anodising, can overcome these shortcomings in aesthetic appearance and utility.

Anodised aluminium has superior atmospheric corrosion resistance and is widely used in the building industry for:

- window frames, door frames, patio doors
- bathroom shower sections and shop fronts
- power trunking and hospital services trunking

The anodising process artificially thickens the passive oxide layer on aluminium. It is an electrochemical process in which an electric current is passed through dilute sulphuric acid and the aluminium article is the anode. A high density of pores leading from the outer surface to the metal interface penetrates the oxide layer formed under these conditions. Immersing the article in boiling water subsequently seals these pores, which are inherent in the anodising process.

Pigments and dyes can be absorbed into the pores before the sealing process to give a wide range of permanent colours. Anodised surfaces with thickness ranging from 5 to $25\,\mu$m are produced for architectural and structural building applications.

- hand rails, door handles and escutcheon plates
- external cladding, hand rails, bridge rails and flagpoles
- signage.

Frequent cleaning is necessary to maintain the durability of anodised aluminium. The finish, degree of soiling, size, shape and location of the installation govern the type of appropriate cleaning. The mildest effective cleaning system should always be used. From the mildest to the most aggressive cleaning systems are as follows:

- plain water
- mild soap or detergent
- solvents (kerosene (paraffin), turpentine and white spirit)
- non-etching chemical cleaners
- wax-based polish cleaner.

More abrasive cleaners (abrasive wax or cleaners) could affect the surface finish and should not be used. The surface should always be rinsed well with plain water and dried with a soft cloth.

Copper alloys

A wide range of copper alloys is used by the construction industry, each with very specific properties and uses. Of these, various grades of copper and the copper–zinc alloys (brass) are the most commonly used. Their significant properties and general use in buildings are summarised in Table 2.6.

Table 2.6 *Use of copper and copper alloys in buildings*

Alloy	Significant properties	Use in buildings
Deoxidised copper	Suitable for brazing	Domestic plumbing tubes
Tough pitch copper	Cannot be brazed	Fully supported roofing sheets and flashing
α Brass (up to 30% zinc)	Can be extensively cold worked	Cold-rolled sheet, wire and tubes
α-β Brass (30–45% zinc)	Can only be hot worked	Castings: valve bodies, pipe fittings
β Brass (45–50% zinc)	Hard and brittle	Brazing rods

Atmospheric corrosion

- Copper and copper alloys tarnish readily, particularly in humid conditions.
- Tarnished copper slowly develops a distinctive patina (a green encrustation on the surface as a result of the reaction of the copper oxides with carbon dioxide and sulphurous compounds in polluted atmospheres), which provides copper roofs with the very distinctive finish seen in many cities of the world.

Galvanic corrosion

- Copper and copper alloys have a high corrosion resistance.
- Copper and copper alloys are cathodic to all other common construction materials, except passivated stainless steel.

Dezincification of brass

- Acidic water, or alkaline water with high chloride content, leaches zinc from brass (particularly α-β brass fittings).
- Dezincification is indicated by a colour change from yellow to copper-red.
- Dezincified brass becomes porous and its mechanical properties significantly impaired.

Further reading

Evans U.R. 1981. *An introduction to metallic corrosion*. Edward Arnold, London.

Fontana M.G. 1986. *Corrosion engineering*. McGraw-Hill, New York.

Uhlig H.H. and Revie R.W. 1985. *Corrosion and corrosion control*. Wiley, New York.

National standard specifications and codes of practice.

Entry web sites

National Association of Corrosion Engineers, USA: http://www.nace.org

Chapter 3

Deterioration of masonry

Contents

Causes of deterioration **52**
Moisture 52
Differential movement 53
    Thermal movement 53
    Irreversible moisture expansion 54
    Dimensional change 55
Environmental impact 55
    Pollution 55
    Frost attack 55
    Fire damage 56
Chemical actions 57
    Lime blowing 57
    Efflorescence 58
    Sulphate attack 59
    Chloride attack 59
    Crystallisation of salts 59
Cement-based mortar and plaster **60**
Natural stone masonry **61**
Durability 61
Surface appearance 61
Frost attack 65
Crystallisation of salts 65
Calcium silicate masonry **66**
Durability 66
Frost action 67
Differential movement 67
Fired clay masonry and paving **67**
Differential movement 67
Lime blowing 68
Frost resistance 68
Attack by salts 69
Cleaning clay brickwork 69

Concrete bricks and blocks 73
Attack by salts 74
Differential movement 74
Further reading 74

Masonry failures occur because of inadequate design or poor workmanship, structures being subjected to forces greater than those allowed for in the design, or the deterioration of masonry materials by factors such as differential movements, environmental impact and chemical actions.

The durability of masonry depends largely on its tensile, compressive and impact strength, hardness, texture and pore structure. Pores can be closed and isolated, or open and interconnected. Some open pores absorb water easily, while others do not. Therefore, it is important to distinguish between the different kinds of porosity.

True porosity: closed pores plus interconnected pores.

Natural absorption: water absorbed on complete immersion in cold water for 24 hours (saturation soaking test).

Total absorption: water absorbed on complete immersion in boiling water for 5 hours.

Saturation coefficient: ratio of natural to total absorption.

Causes of deterioration

Moisture

Water is an essential factor in the majority of conditions causing deterioration of masonry. For example:

- water leaches salts from the surroundings and transports them by percolation into porous materials or deposits them on the surface of such materials
- water dissolves pollutants to form harmful acids
- near-saturation conditions are needed for frost action to occur

- moisture is needed for irreversible moisture expansion or lime blowing.

In those instances where moisture plays a role, the first step in any remedial action is to rectify the conditions leading to the ingress of water into the structure. This may entail design changes or structural adaptations, which could require involved procedures that fall outside the scope of this guide.

Differential movement

Thermal movement
There is a considerable variation in the unrestrained thermal expansion of masonry materials, both between and within the different types, as shown in Table 3.1.

- Thermal gradients will generate stresses in materials. However, diurnal temperature fluctuations, even the sudden temperature change when rain falls onto a surface that was exposed to direct sunshine, are rarely sufficient to result in spalling of building materials as such.
- On the other hand, the difference in thermal expansion coefficients between different building materials could be sufficient to cause the failure of composite systems, such as a ceramic cladding on a concrete substrate. If effective provision is not made for differential thermal movement, the cladding could buckle, crack or detach from its substrate.

See also:
  Concrete, Thermal expansion, p. 30
  Tiling, Thermal movement, p. 82

Table 3.1 *Linear thermal expansions of selected masonry and cladding materials*

Material	Expansion (mm/m per 20°C)
Limestone and marble	0.04–0.14
Granite, sandstone and slate	0.08–0.12
Fired clay bricks and tiles	0.12–0.18
Cement products	0.22–0.24
Calcium silicate bricks	0.16–0.28

Irreversible moisture expansion

All porous materials (not only ceramics) expand to some extent when they absorb water. This kind of expansion is fully recovered when the material dries out. Although this type of reversible change is very small, it may become significant because of its nature over the lifespan of a material. However, most silicate ceramic materials expand irreversibly when they get wet. In other words, the expansion remains after ordinary drying. This irreversible 'moisture expansion' is caused by chemical reactions between the moisture and the glassy phases in the ceramic body. These reactions start as soon as the ceramic leaves the dry atmosphere of the kiln and comes into contact with moisture. The resulting expansion is rapid initially, with approximately half of the potential expansion developing during the first 10–12 weeks of exposure. The expansion will continue for many years, although at an ever-decreasing rate.

- The moisture expansion of some ceramic bodies can be considerable. Linear expansions of 0.1% (1.0 mm/m) after 5 years of normal ambient exposure are not unusual, while expansions as high as 0.2% have been recorded.
- The firing temperature and the composition of the raw materials greatly influence the magnitude of the irreversible moisture expansion of a ceramic.

Bodies with high porosity and/or large amounts of glassy phase have a high potential for moisture expansion. Low-fired bodies (high porosity but little glassy phase) and high-fired bodies (glassy but dense) thus tend to have lower moisture expansions than do bodies fired at intermediate temperatures. However, the firing temperature determines the colour and texture of ceramics, and aesthetics may dictate the selection of ceramics fired at intermediate temperatures.

The main factor that determines the magnitude of the expansion is the composition of the raw materials. Research has proved that the alkali metals sodium and potassium are the main culprits. It was also found that the moisture expansion could be dramatically reduced if the alkali metals were replaced or diluted with the alkaline earth elements calcium and magnesium. This is achieved by adding very finely ground limestone, dolomite, talc (a magnesium silicate) or wollastonite (a calcium silicate) to the raw material blend. The adjustment of the composition of the raw materials is the best approach to minimise the problem of irreversible moisture expansion of silicate ceramics.

- High humidity and ambient temperature initially accelerate the expansion, but it seems that the ultimate expansion is independent of the exposure humidity or temperature.
- Physically restraining the ceramic will not prevent it from wanting to expand.

See also:
   Tiling, Irreversible moisture expansion of tiles, p. 82
   Tiling, Failure of tiled systems, p. 83

Dimensional change
Products, such as cement-based products, shrink when they dry and creep under load. Most of the drying shrinkage occurs during the first few weeks of making, but concrete and cement products could shrink linearly another 0.02% over the next year or two. Creep obviously depends on the stress level, but the long-term creep of fully loaded concrete pillars could be as much as 0.2%.

These dimensional changes in concrete materials could cause considerable stresses in a structure, particularly when expanding materials, such as certain ceramic products, form integral parts of the structure.

See also:
   Concrete, Shrinkage and creep, p. 30
   Tiling, Differential movement, p. 82
   Tiling, Failure of tiled systems, p. 83

Environmental impact

Pollution
The most noticeable aspect of air pollution is the accumulation of grime, particularly in areas protected from driving rain. Much more serious in polluted environments is the reaction of the sulphurous and nitrous rainwater with carbonate in calcareous masonry elements. The presence of algae, fungi and lichens, particularly on rough surfaces, aggravates the attack, as some of these organisms additionally promote the conversion of sulphur and nitrogen from the polluted atmosphere to sulphurous and nitrous acids. They also serve as traps for aggressive substances and prevent them from being washed away by rain.

If any calcium hydroxide formed on the surface of stone cladding is not washed away immediately it could carbonate, and an unsightly hard skin (usually discoloured by air-borne pollutants) is formed that is resistant to removal.

Frost attack
Frost damage generally occurs when wet porous materials are exposed to

55

frequent freeze–thaw cycles. The simplistic explanation is that when the water in the pores freezes, the resulting 9% volume expansion causes the disintegration of the material. However, the mechanism of frost damage is not simple and the pore structure has a decisive influence on the thermodynamics of ice formation, thus controlling the structure of the ice crystals and the microstructure of the ice mass, which in turn determines the extent of the damage.

- The material must be permeable. The pores must be interconnected and big enough to be filled easily by water. The extent to which pores will be filled with water when the material is immersed or exposed to driving rain is reflected by its saturation coefficient. Values for porous building materials range from 0.4 to 1.0, and there are indications that materials with high saturation coefficient values are more at risk of damage due to frost action, but the correlation is not high.
- The pore size distribution is the determining factor. Water in pores is subjected to surface tension and capillary forces and these forces depress the freezing temperature of water. The forces increase with decreasing pore size, with the result that water in large pores freezes at about 0°C, while water in pores <0.5 μm does not freeze even at very low temperatures. Therefore, the water in the larger pores freezes first, while the water in smaller pores remains liquid. As the temperature drops, water in the smaller pores feeds the larger pores, increasing the pressure within them. When the pressure exceeds the tensile strength of the material, the material starts to spall. This is also the mechanism through which ice crystals can be extruded from pores or cracks near the surface.
- Frost damage is caused by cycles of freezing and thawing. It is cumulative and leads to flaking of the surface layers or even to total crumbling of the material. Even slight damage is usually unacceptable, because the general surface appearance is impaired.

Predicting frost resistance from material properties (strength, pore characteristics and saturation coefficient) or accelerated freezing tests is not very successful, and thus materials are classed according to actual exposure tests or their performance track record.

See also:
  Tiling, Frost attack, p. 78

Fire damage
High temperatures break down hydraulic cementitious compounds. The strength of cement products is not severely affected by temperatures below 250°C, but drops to 80% at about 450°C and is only 50% at about 650°C. On cooling, there is a further loss in strength.

- The very rapid temperature changes experienced in a fire situation could cause cracking and spalling of ceramic materials.
- If the temperature exceeds the original firing temperature of the ceramic, further physicochemical changes will occur.

See also:
Concrete, Effect of high temperatures, p. 28

Chemical actions

Lime blowing

> When heated to temperatures above 600°C, limestone and dolomite particles start to decompose to quicklime (CaO) and periclase (MgO), respectively, and carbon dioxide. When such particles come into contact with moisture they start to react with the water to form calcium or magnesium hydroxide. This change results in a marked increase in volume of the hydrating particle and, if the particles are enclosed in a strong matrix, large crystallisation forces are created. If the hydrating particle is large enough and close to the surface, a piece of the matrix is popped off, leaving a characteristic crater with powdery product at the bottom of the pit.

Limestone is a common impurity in clay materials used in the manufacture of fired clay products. During the clay preparation step in the brick manufacturing process, any large lumps of limestone will be crushed to small fragments, of the order of a millimetre in size. It is the decomposition during firing of these limestone fragments and subsequent hydration of the quicklime nodules that cause lime blowing to occur in fired clay products. The hydration reaction cannot be stopped as long as moisture is present. Lime blowing may become apparent within days of construction or only after some months or even years, depending on the temperature at which the ceramic has been fired, the humidity to which the ceramic is exposed and the depth of the particle from the surface. Note that reasonable amounts of finely divided calcium carbonate will fully react with the clay minerals during the firing process to form stable calcium aluminosilicates.

Fully hydrated building lime improves the workability of cement mortars and plasters. Poor-quality building lime may still contain some large particles of quick lime that were not slaked completely during manufacture. The continued slaking of these large particles could cause subsequent lime blowing in the mortar or plaster.

Coal ash and clinker intended for making cement products, such as concrete masonry units, should be carefully tested for lime nodules, to avoid lime blowing or even complete destruction of the products.

See also:
Tiling, Lime blowing, p. 76

Efflorescence
Some fired clay products contain soluble salts (mainly various types of sulphates formed during the firing process) that migrate to the surface on repeated wetting and drying of the brickwork. Masonry may also be contaminated by water percolating through adjacent sulphate-containing concrete, mortar backfill, or by sulphate-containing effluent. The deposits that form on the surface of masonry or in patches on rendered brickwork are usually white, but could be green, yellow or brown. This kind of salt deposit is known as *efflorescence*. It is unsightly and could be detrimental to under-burnt bricks, but not well-burnt bricks. Layers of plaster and paint, however, will be lifted, particularly if the paint is impermeable.

The sources of soluble sulphate in fired clay products are:

- the dehydration during firing of any gypsum ($CaSO_4 \cdot 2H_2O$) in the raw materials to the anhydrite $CaSO_4$
- the reaction of lime in the raw materials with sulphurous kiln gases during firing to form $CaSO_4$
- the reaction of lime in the raw materials with sulphur-bearing substances in solid fuels mixed into the clay body.

Small amounts of vanadium-bearing minerals are present in most brick-making clays. During the firing process these minerals may be converted to soluble forms. Vanadium efflorescence or staining appears as a light greenish stain that is more noticeable on light-coloured bricks than on dark-coloured bricks, but normally weathers away in time. Vanadium efflorescence should not be confused with micro-organism growths. Manganese dioxide, sometimes used as a pigment in certain face bricks, could cause the appearance of dark brown to violet stains on the bricks. Although unsightly, vanadium or manganese efflorescence is not detrimental to brickwork.

Potassium carbonate efflorescence may develop on calcium silicate bricks. The lime used in their manufacture reacts with the potassium in

> There are standardised procedures for testing masonry
> units for efflorescence. Specimens are placed in 25 mm deep
> clean, salt-free water and left in a well-ventilated area at
> 20–30°C until all the water evaporates. The process is
> repeated and the specimens then classified into five
> categories according to the extent of the salt deposits
> formed on the specimens.

feldspar or mica if present in the fine aggregate, forming potassium hydroxide. This salt percolates to the surface, where it carbonates to form a fluffy efflorescence. It is, although unsightly, fairly harmless and washes off easily because of its high solubility.

Sulphate attack
Sulphate solutions react with hydrated lime and calcium aluminates and ferrites in cement in set mortars, resulting in disruptive expansion. The reaction requires high moisture levels. The sources of sulphate are:

- soluble sodium, calcium and magnesium sulphates formed in building ceramics during the firing process
- soluble sulphate formed by reaction between carbonate in masonry elements and sulphurous polluted rainwater
- soluble sulphate in the surrounding soil
- sulphate-bearing solutions, such as domestic and industrial effluent.

See also:
  Concrete, Sulphates, p. 26

Chloride attack
Chlorides in significant quantities break down the bond in calcium silicate products. The deterioration mechanism is not completely clear, but the solubility of calcium carbonate (which becomes an important binding phase in the carbonated regions of calcium silicate products) increases significantly in salt-laden environments (sea spray, salt-laden mist or de-icing slush). The breakdown of the binding phases results in an increase in friability of the calcium silicate product.

Crystallisation of salts
Porous materials disrupt when salt solutions in the pores crystallise and the process is accompanied by an increase in volume. Evaporation causes crystallisation in the pores just below the surface, and repeated cycles of

leaching and drying cause the outer layers of the material to disintegrate slowly. Some salts are hygroscopic and take up water when the humidity is high, resulting in a further increase in volume.

The appearance of this kind of disruption, called *crypto-efflorescence*, is very similar to that caused by frost action, but the presence of salt deposits in the failure zone is usually very noticeable. There are various sources of soluble salts:

- soluble salts in the mix materials of cement products
- sodium chloride solutions, such as seawater or road de-icing run-off
- brackish groundwater and salt-laden permeating solutions
- sulphates and nitrates formed by reaction between polluted rain and calcareous phases.

The durability of porous building materials against salt attack is evaluated using the crystallisation test. Test specimens are subjected to 15 cycles of soaking in a sodium sulphate solution and drying. The condition of the specimens is ranked visually using a six-point scale from A (high resistance) to F (low resistance).

See also:
Concrete, Chemical attack on concrete, p. 25

Cement-based mortar and plaster

Mortar refers to the bedding material that binds masonry units, while *plaster* refers broadly to the coating material on exterior and interior walls, ceilings and floors. In general, but not universally, the term *rendering* is applied to cement-based coatings for water-resistant exterior surfaces, while the term *plaster* is applied to materials that are used for providing durable, smooth and easily decorated interior wall or ceiling finishes. Plasters in this sense are either cement–lime–sand mixtures or gypsum materials. Cement-based floor finishes are referred to as *screeds*.

More than one agent, as summarised in Table 3.2, frequently cause failure of cement-based mortars and plasters, or it is the consequence of a different failure. It should be stressed that where water penetration into a building structure plays a role, repair work would be unsuccessful if the structure is not waterproofed before repair work is carried out.

Cement–sand mortar and plaster range from high strength (mortars for multi-storey load-bearing walls or plasters exposed to aggressive conditions) to low strength (mortars for single-storey brickwork or plasters for very weak brickwork). Straight mixes of Portland cement and sand are too harsh and difficult to work, and thus plasticisers, such as building lime, and air entrainers are incorporated in the mix. Alternatively, special masonry cement may be used, which is a blend of common cement, building lime, ground limestone, extenders and/or air entrainers.

High cement contents increase strength and improve resistance to weathering and penetration of damp, but increase the tendency for shrinkage.

The binder in gypsum plasters is plaster of Paris, which is produced by the controlled dehydration of gypsum (calcium sulphate dihydrate). Plaster of Paris rapidly sets when mixed with water, changing back to gypsum. Gypsum plasters are supplied in different setting-time classes. The setting rate of plaster of Paris depends on the dehydration temperature, but can be controlled by additions or retarders (e.g. keratin) or accelerators (e.g. alum). Gypsum plasters are non-shrinking, but are not suitable for exterior use because gypsum is slightly soluble in water.

Natural stone masonry

A large variety of natural stone is used in building, ranging from hard to easily worked, dense to porous, and with many colours, shades and textures.

Durability

There is considerable variation in properties and durability, not only between rock groups, but also within a specific type of rock, as summarised in Table 3.3. The main agents of deterioration are atmospheric pollution, frost action and salt crystallisation.

Surface appearance

The most noticeable effect of atmospheric pollution is the deterioration in appearance. Grime and micro-organisms accumulate particularly on surfaces protected from rain. Staining, sometimes accompanied by efflorescence, of the lower courses of porous stone ashlar could also result

Table 3.2 *Causes and repair of failure of mortar and plaster*

Cause of failure	Repair
Crumbling and powdering low-strength mortar or plaster	
Mortar too lean	If structural integrity has been compromised, rebuild the affected sections
Frost action	Rectify the conditions causing the masonry to get soaking wet during frost conditions
	Rake out affected mortar, re-point with appropriate mortar
Sulphate attack	Rectify conditions causing wet condition of brickwork
	Rake out affected mortar, re-point with sulphate-resisting mortar
Fire damage	Damage is usually restricted in depth
	Rake out affected mortar and re-point with appropriate mortar, or re-plaster after removing all old plaster
Plaster that has developed fine cracks in a random pattern	
Plaster mix too strong, or sand too fine or clayey, thus plaster has high drying shrinkage	Cracks allow rainwater penetration, which creates conditions for sulphate attack
	Cover and seal fine cracks with cement paint or fibre-filled polymeric paint

Plaster shows a prominent horizontal crack pattern, following the brick courses

Sulphate attack of mortar, with vertical expansion of brickwork	Rectify wet condition of brickwork
	Seal cracks and decorate with flexible paint
	Re-plaster or install cladding in extreme cases

Rendering detached from brickwork or concrete substrate

Efflorescence, where salts have crystallised at the interface between the plaster and the substrate. The presence of salts is generally easily detected	Rectify conditions that allow salt-laden water to percolate into the substrate
	Chip away all plaster in the affected area, roughen the substrate to provide a mechanical bond and re-plaster

Craters pock marking plastered masonry

Lime blowing	Any quicklime nodules in plaster are generally small and hydrate fairly rapidly, but the occurrence of lime blowing could continue for a considerable period after construction. Repair will be a continuing action. Fill in craters, or surface coat (skim), and then decorate
	Gypsum skim coats must be applied only to plasters made with clean, well-graded sand, which was first allowed to dry and shrink adequately
	Lime blowing in cement products, such as concrete blocks made with clinker containing large quicklime nodules, could continue for years

Table 3.3 *General durability of natural stone masonry*

Class	Group examples	Durability
Igneous	Granite (coarse grained)	Highly durable
	Dolerite (medium grained)	Hard: slow surface wear
	Basalt (fine grained)	Impermeable: highly resistant to environmental deterioration
		Surface discoloration by pollution, weathering and organic growth
Metamorphic	Marble (massive)	Very durable
	Slate (strongly bedded)	Low to zero porosity: generally very suitable for exposure to severe environmental conditions
		Good resistance to surface wear
Sedimentary	Sandstone (siliceous)	Variable durability
	Limestone (calcareous)	Generally porous with a relatively rough surface, and variable resistance to salt and frost attack
	Shale (argillaceous)	
		Deterioration mainly caused by atmospheric pollution

from splash-up at pavement level or if a defective damp proof course allows rising moisture.

If certain calcium salts are deposited on the surface and are not washed away immediately by either rain or a cleaning action, an unsightly hard skin is formed, which is resistant to removal.

- Polished dense stone, such as basalt and granite, is readily cleaned by washing or spraying with water containing mild detergent and light brushing, followed by buffing. Care is needed around embedded ferrous metal fittings or timber.
- Highly polished marble should not be used externally, as it does not keep its polished appearance.
- Porous stone, such as sandstone and some limestone, is less easy to clean and generally needs professional services. High-pressure water jets, wet or dry grit blasting or abrasive brushes may be required.
- Stone, except calcareous stone, can be cleaned by ammonium bifluoride or diluted hydrofluoric acid. Only professional cleaners

should carry out this method, as these chemicals are extremely dangerous to handle and great care is needed to avoid damage to adjacent materials, particularly glass. Caustic soda and soda ash are damaging and must not be used on stone.

Apart from maintaining good appearance, regular cleaning of stonework and repair of joints minimises the possibility of other defects developing unnoticed.

Frost attack

The resistance of natural stone to frost action is variable, because the porosity and pore structure of these materials may differ considerably from unit to unit. Some individual units in natural stone masonry or paving, therefore, commonly show severe deterioration, while others are apparently immune to frost action. The possibility that chemical attack or crystallisation of salts are contributory factors must always be considered.

- Prevent saturation wetting of stone.
- Protect stone from freezing–thawing conditions.
- Use only stone with a proven record of frost resistance for any repair or reconstruction work.

See also:
  Concrete, Frost attack, p. 28
  Tiling, Frost attack, p. 78

Crystallisation of salts

The main source of salts causing disruption of the surface of stone is the reaction between polluted rain and calcareous phases in the stone. Siliceous sandstones and most slates are rarely destructively affected by atmospheric pollution. Although all limestones are attacked to some extent, their resistance is very variable and some show good durability even in the worst environments. Calcareous sandstones are, on the whole, liable to be badly attacked by environmental pollution. The resistance of porous stone to salt attack is evaluated by means of the crystallisation test.

The crystallisation or hygroscopic nature of sulphates and nitrates in pores near the surface during cycles of wetting and drying causes flaking and spalling of the surface, known as *contour scaling*. This disruption is more pronounced if the bedding of the stone is parallel with the surface. In appearance this disruption is very similar to that caused by frost action.

- Cleaning by specialist services may arrest the process, if the deterioration is detected early. Stone prone to contour scaling should thus be cleaned regularly.

- Micro-organisms on the surface could aggravate the disruption of the stone by chemical attack. These organisms convert sulphur and nitrogen from atmospheric pollution to sulphurous and nitrous acids, intensifying the attack on the stone. They also trap aggressive agents, keeping these in contact with the stone, and prevent their easy removal by rain.
- Replacement of a substantial part or of the whole of the stone may be required if the deterioration is extensive. Great care should be taken in the selection of replacement stone, particularly to ensure equivalent rates of weathering between the remainder and replacement stone.
- Injudicious use of surface sealants to stop flaking could have disastrous consequences, as such treatment could actually aggravate the flaking.

See also:
   Concrete, Chemical attack on concrete, p. 25
   Tiling, Salt attack, p. 78

Calcium silicate masonry

Provided that the appropriate class of material is used, calcium silicate products are adequate for almost all purposes and are compatible with all materials generally used for damp-proving, flashings, wall ties, metal fasteners, etc. They are rarely destructively affected by atmospheric pollution.

A moist mixture of hydrated lime, sand or crushed flint is pressed and then steamed at pressures up to 2.0 MPa for 8 hours in an autoclave. During this process the lime reacts with the silica surface to form a strong calcium silicate bond, leaving no free lime.

Calcium silicate bricks are classed according to their mean wet crushing strengths, ranging from 14.0 MPa to 48.5 MPa. During use the calcium silicate binder gradually reacts with carbon dioxide from the atmosphere to form calcium carbonate. The bricks slowly gain in strength and hardness and finally resemble natural calcareous sandstone.

Durability

Calcium silicate bricks exposed to salt spray (sea spray, salt-laden mist or de-icing slush) may rapidly become friable. Erosion rates of affected surfaces exposed to heavy rain or high winds may be 5 mm or even more per season.

- If the conditions are severe, the exposed brickwork will have to be replaced. Rendering the brickwork may not be successful, because of the difficulty of completely removing the affected skin of the units. Alternatively, the brickwork may be protected from the elements by installing cladding, but the fixing must be to unaffected material.
- In situations where the deterioration is relatively slow, surface treatment with silicone or linseed oil may retard, but not eliminate, the deterioration. However, surface sealers must be used only with circumspection.
- Potassium carbonate efflorescence may develop on calcium silicate bricks. It is fairly harmless, although unsightly, and washes off easily.

Frost action

Resistance to frost action is related to the strength class of the calcium silicate bricks. However, even high-strength class bricks are liable to damage by frost action if chlorides contaminate them. If this situation arises, the affected calcium silicate units should be replaced with other materials with a proven record of frost resistance.

Differential movement

The thermal movement of calcium silicate bricks is similar to that of concrete, but higher than that of fired clay products. The drying shrinkage of calcium silicate bricks is about 0.035–0.040%, and carbonation shrinkage may also occur. Crack development can be expected at junctions with other construction materials such as concrete columns, beams, slabs and lintels, or fired clay brickwork.

The low suction (ability to draw in water) of calcium silicate bricks adversely affects the bond strength with mortar. Mortar made of fine clayey sand also results in bricks becoming loose and creep out of place, particularly in free-standing walls and parapet walls.

- Little can be done to rectify poor bonding between the bricks and mortar, although raking and deep pointing could help somewhat.
- Appropriate movement joints should be installed and maintained.

Fired clay masonry and paving

Differential movement

Structural damage to masonry could be caused by differences in thermal expansion, irreversible moisture expansion, drying shrinkage and creep of the different materials employed.

A very large range of fired clay masonry units is produced (solid and perforated bricks, hollow blocks, ceramic planks, pavers, etc.) for a variety of applications: rendered or plastered (commons, stocks), unrendered (facing) and aesthetic (rustic, rock-faced, clinkered). The appearance of the products (colour and texture) is controlled by the choice of raw materials and manufacturing conditions.

Durability depends mainly on the nominal compressive strengths, which ranges from 3.3 MPa for non-load-bearing applications, to 12.0 MPa for facing bricks, to 17.0 MPa for high-load-bearing applications. The porosity of the products decreases with increasing strength.

The only reason for structural damage to occur is the failure to make adequate provision for dimensional changes and the resultant differential movements by the correct design of movement joints and/or their proper maintenance.

- Frequent inspection and maintenance of movement joints are vital, because early warning of excessive differential movements may be given by the expression of sealants from these joints.
- Partial or total demolition and reconstruction is usually the only option available if structural damage has occurred.

Lime blowing

The craters formed by lime blowing may start to appear a few days after the bricks or pavers have been laid, or only after some months, depending on the temperature at which the ceramic has been fired and the humidity to which the ceramic is exposed.

- Limited surface damage to roughly textured bricks may not be very noticeable, but the appearance of smooth-textured bricks may become unacceptable.
- Lime blowing could continue for a long time after construction. Rendering or plastering should thus be postponed until the incidence of lime blowing has dropped sufficiently so that frequent crater repair is not required.
- Severe lime blowing could compromise the structural integrity of the brickwork, making demolition and rebuilding necessary.

Frost resistance

The damage caused by frost action is cumulative and will continue unabated to complete destruction while the brickwork or paving remains saturated during periods of cyclic freezing and thawing. The possibility that chemical attack or crystallisation of salts are contributory factors must always be considered. Early and prompt remedial action is vital, because even slight damage by flaking of the surface layers may already be unacceptable.

- Rectify all conditions causing the saturation of the brickwork.
- Waterproofing treatment may be helpful, if the source of water is external (rain or melting snow). Be aware that surface sealers could aggravate the situation.
- Protect the brickwork from freeze–thaw conditions by rendering or cladding with weather screens.
- Use only bricks and pavers with a proven record of frost resistance for any repair or reconstruction work.

Attack by salts

Deterioration of masonry by soluble salts is fairly common, particularly in retaining walls, where these salts may contaminate brickwork. Crystallisation of the salts, or hygroscopic action, causes the brickwork to disintegrate.

- Prevent, if possible, the ingress of salt-laden water into the brickwork.
- Use dense and strong masonry units for repair and reconstruction.

Cleaning clay brickwork

The reasons for the degradation in appearance of brickwork could be actions during construction (e.g. mortar smears), processes inherent to the bricks (e.g. efflorescence and staining) or actions of outside agencies (e.g. graffiti). These kinds of blemishes do not lead to structural failure.

Appropriate cleaning procedures are summarised in Table 3.4. It is always highly advisable to contact representatives of the brick manufacturer for confirmation as to the appropriate cleaning action for a particular occurrence.

- Do not use concentrated acid, as it will attack the mortar and could cause acid staining ('brick burn') of light-coloured bricks. Hydrochloric acid should always be diluted to one part acid to ten parts water.
- Chlorine vapours emanate from hydrochloric acid. These vapours will cause severe rusting of any exposed steel fittings, and thus interior areas must be very well ventilated during and after the cleaning operation. Direct contact of steel with hydrochloric acid should also be avoided, and treated areas should be thoroughly washed to remove all chlorides.
- Do not use sulphuric acid (battery acid) instead of hydrochloric acid.
- It is important that the surface of the masonry is saturated with water

Table 3.4 *Cleaning clay brickwork*

Defect	Treatment	Comments
Fresh mortar smears (<1 day old)	Scrub with hard bristle brush and clean water. Do not use steel brushes, as iron rubbed off will cause iron rust staining	Avoid using more aggressive methods (dilute acids or cleaning agents)
Hardened mortar (>1 day old; needs more aggressive treatment)	First saturate masonry with water. Scrub with 1:10 diluted hydrochloric acid. Alternatively, use proprietary cleaning agents (weak solutions of hydrochloric acid or oxalic acid plus modifiers). Wash down very well with clean water	Do not use concentrated acid. Hydrochloric acid is also known as muriatic acid or spirits of salts
Lime deposits (leached free lime from mortar or concrete)	Scrub with clean water and, if not successful, first saturate masonry with water and scrub with dilute hydrochloric acid. Wash down well with clean water	Distinguish lime deposits from sulphate efflorescence
Sulphate efflorescence (leached sulphates from bricks, mortar or soils)	Exterior walls: hose down, repeat a few times if necessary. Internal walls: allow to dry completel and then brush salts off	Distinguish from lime deposits. Do not treat with alkaline or acidic agents. Recurring efflorescence indicates continued migration of water
Vanadium efflorescence (thin greenish stain)	Accelerated removal is problematic. Try treating dry stain with: (a) diluted oxalic acid (20–40 g/l); neutralise with dilute washing soda	Stain tends to weather away with time. Do not confuse stain with algae growth. Oxalic acid is extremely toxic

	(b) diluted hypochloride (100 g pool chlorine/l)	Do not use other acids, as stabilised salts are formed
	(c) diluted caustic soda or washing soda (100 g/l)	
	(d) acetic acid or hydrogen peroxide	
Manganese efflorescence (dark brown to violet stain)	Easily removed by proprietary cleaners (based on hydrochloric acid and sodium fluoride)	Wash down well but do not neutralise with alkalis
Iron rust staining (yellow-rust colour)	*Rust stains:*	Oxalic acid is extremely toxic
	(a) wash with (50 g oxalic acid + 20 g sodium fluoride + 15 g citric acid)/l	Do not use steel brushes, as the iron rubbed off will cause further iron rust staining
	(b) neutralise with 50 g/l bicarbonate of soda	
	Light staining (walls discoloured by irrigation water):	
	(a) use diluted phosphoric acid (1:4)	
	(b) wash down with clean water	
Fresh emulsion (water-based) paint	Scrub down with large amounts of clean water, or use water jetting or grit blasting	Cleaning is seldom completely successful
Oil-based paint and old paint	*Mechanical treatments:*	Mechanical treatments are highly abrasive
	(a) sand blasting, power grinding or water jetting	Some chemical treatments require specialist supervision
	Chemical treatment:	Laser treatment is very expensive and slow
	(a) standard paint removers on smooth surfaces	

Contd

Table 3.4 *Contd*

Defect	Treatment	Comments
Oil-based paint and old paint – *contd*	(b) apply caustic poultices (caustic soda 300 g/l) mixed with inert filler (diatomaceous earth, flour, methylcellulose), leave for 2 days, hose down with a high-pressure water jet	
	(c) gels containing methylene dichloride, isoamyl acetate, ethylmethylketone, etc., are used by specialist cleaners	
	Laser treatment: paint is evaporated by high-power lasers	
Oil, bitumen, tar	Bitumen and tar: requires specialist services	Do not wet bricks
	Solvents for mineral oils: petrol, benzene, naphtha	Complete removal is very difficult and treatment may spread the contaminant
	Solvents for vegetable oils: methylated spirits, turpentine, trichloroethylene	Weathering slowly breaks down oil contamination
Micro-organic growths (fungi, moulds, lichen, moss)	Wash with:	Some chemicals are harmful or toxic to humans, animals and plants
	(a) copper sulphate (15 g/l)	Rectify the conditions promoting the growth of the organisms
	(b) sodium hypochlorite (full strength)	
	(c) sodium pentachlorphenate (50% solution)	
	(d) formaldehyde (125 ml/l)	
Grime	Scrub down with water and detergent. For persistent grime use high-pressure water jets, wet or dry grit blasting, or abrasive brushes	Use the mildest method to avoid unnecessary surface abrasion

to limit the uptake of cleaning agents, especially acids that could subsequently damage the bedding mortar.

- Verify that efflorescence deposits are not leached lime from the mortar or adjacent concrete.
- Efflorescence is unsightly and could be detrimental to under-burnt bricks, but not well-burnt bricks. Layers of paint, however, will be lifted, particularly if the paint is impermeable. Painting of walls showing signs of efflorescence should be postponed until the walls have dried out completely and all the salt deposits have been brushed off.
- A warning must be sounded against the use of sealants or water repellents to get rid of efflorescence, because of the risk of spalling of the brick surface due to the crystallisation of the salt behind the sealant or impregnated layer.
- If efflorescence persists, it indicates continuous migration of water through the brickwork (a design or construction failure) or that sources of soluble salts other than the bricks are involved (e.g. contaminated mortar, concrete or fill). Conditions allowing moisture movements must be rectified.

Safety precautions

- Cleaning agents are generally aggressive and toxic. Many cleaning systems are potentially hazardous to humans, animals or plants. Note that oxalic acid is extremely toxic.
- Protective clothing, eye protection and, in some cases, respirators must be used during cleaning operations.

Concrete bricks and blocks

- A damp concrete mix made with fine aggregate is pressed or vibrated into moulds. The moulded units are then cured either in air or in steam chambers.
- The bricks are in standard sizes and are sometimes coloured. The higher strength classes may be used as facing bricks.
- The blocks are typically 390 mm long, 140 or 190 mm wide and 90 mm high, with two large vertical cavities. Walls constructed with blocks are generally rendered.

Concrete bricks and blocks are inexpensive masonry units and they are used extensively in low-cost projects.

Attack by salts

Crystallisation of soluble salts, or hygroscopic action, could cause concrete bricks and blocks to disintegrate. However, these materials do not suffer from chloride attack in the same way as calcium silicate units.

Differential movement

The drying shrinkage of concrete bricks and blocks during the first 4 weeks after manufacture is about 0.02%, with a further 0.02% shrinkage over the first year or two. Because of the large size of the blocks, this drying shrinkage could result in a parting between the mortar and block, compromising the resistance of the wall to rain penetration.

* Do not use freshly made blocks, but allow curing for some weeks.
* Delay rendering as long as possible and use waterproofing agents in the plaster.

See also:
   Concrete, Dimensional change, p. 29
   Tiling, Salt attack, p. 78

Further reading

Addis B.J. and Owens G. (eds) 2001. *Fulton's concrete technology*, 8th edn. Cement & Concrete Institute, Midrand.
Clews F.H. 1969. *Heavy clay technology*. Academic Press, London.
May H. 1997. *Technical Guide No. 1–4*. South African Clay Brick Association, Cape Town.
National standard specifications and code of practices.

Chapter 4

Tiling failures

Contents

Material failure **76**
Crazing of glazed ceramics 76
Lime blowing 76
Frost attack 78
Salt attack 78
Tile adhesive failure 78
    Inappropriate adhesive 79
    Attack by aggressive chemicals 80
    Poor workmanship 80
Preventive and remedial actions 80
System failure of tiled floors and walls **81**
Differential movement 82
    Thermal movement 82
    Irreversible moisture expansion of tiles 82
    Drying shrinkage and creep of cement substrate 83
Failure of tiled systems 83
Preventive measures 84
    Differential movement 84
    Movement joints 85
    Workmanship 85
Further reading **85**

One of the major causes of the failure of tiling systems is poor workmanship, particularly incomplete bedding of the tiles in the adhesive, which allows puncturing and other kinds of impact damage to occur. Laying tiles on areas of adhesive where the open time has been exceeded, resulting in poor adhesion of the tiles, is another common example of poor workmanship. Poor design (e.g. tiling across movement joints, tight butting against projections) is also a general cause of tiling failure. Therefore, recognition or elimination of poor design and workmanship as the root cause should always be the first step in an investigation of tiling failure.

Some tiling failures manifest themselves quite soon after the tiles have been fixed, while other failures may take many months or even years to become apparent. Two types of failure of tiled areas that were initially sound can be distinguished: material failure and system failure.

Material failure

The causes of the progressive deterioration of the surface of tiles, the body of tiles and the adhesive are summarised in Table 4.1.

Crazing of glazed ceramics

A common, but avoidable, failure of glazed ceramics is crazing. In the case of tiles, fine hairline cracks in a roughly concentric pattern may start to appear on the surface of the tiles, many months or even years after the tiles have been fixed. The cause of this kind of failure is the irreversible moisture expansion of the tile body that places the glaze layer in tension. When the resulting tensile strain exceeds a critical value, the glaze layer ruptures and fine cracks are formed. These cracks become noticeable when they start to fill with dirt.

Artist potters sometimes deliberately cause crazing of glazed articles for aesthetic reasons. When glazes with low thermal expansions are used on bodies with higher thermal expansions, the glaze layer will be put into tension on cooling from glost firing. If the tension is sufficiently high, the glaze layer will fail by forming fine hairline cracks. By adjusting the amount of thermal expansion mismatch, the degree (but not the detailed pattern) of crazing can be controlled.

See also:
Masonry, Irreversible moisture expansion, p. 54

Lime blowing

Lime blowing manifests itself as small craters on the surface of tiles, usually with a tiny nodule of white powder at the bottom of the pits. This

Table 4.1 *Identification of material failure in tiling systems*

Defect	Appearance	Cause
Crazing of glazed tiles	Glaze develops fine hairline cracks in a roughly concentric pattern	Irreversible moisture expansion of tiles
Lime blowing	Appearance of small craters with a tiny nodule of white powder at the bottom of a pit	Hydration of particles of quicklime in the tile body (quicklime is formed during firing from fragments of limestone in the raw material)
Frost damage	Progressive disintegration or flaking of the tile surface exposed to cyclic freeze–thaw conditions	Repeated expansion of adsorbed water in tile pores when frozen
Abrasion	Progressive wearing away of the surface in heavy traffic conditions	Low inherent strength and fracture toughness of the tile
Adhesive failure	De-bonding of tiles not associated with dimensional change	Incorrect adhesive used

defect is particularly unacceptable in glazed tiles with strongly contrasting body and glaze colour.

Large limestone particles in the tile body decompose to quicklime (CaO) during the firing process. The subsequent hydration of these particles results in a marked increase in volume, creating large crystallisation forces. If the hydrating particle is near the surface of the tile, a small piece is popped off.

Note that reasonable amounts of finely divided calcium carbonate in the clay body will fully react with the clay minerals during the firing process to form unreactive calcium aluminosilicates.

The craters may start to appear a few days after the tiles have been laid, or only after some months or even years, depending on the temperature at which the tiles have been fired, the humidity to which the ceramic is exposed or the depth of the particle from the surface.

See also:
Masonry, Lime blowing, p. 57

Masonry, Lime blowing, p. 57

Frost attack

The surface of porous tiles exposed to cyclic freezing and thawing while saturated may progressively disintegrate. In certain instances the repeated freezing of the pore water causes internal crack formation. The volume of the tile increases and this cumulative expansion is known as *frost heave*.

There are several conditions necessary for frost damage to develop:

- the tiles must have high levels (i.e. near to saturation point) of adsorbed water
- the tiles must have high permeability (i.e. the pore structure must be interconnected)
- the tiles must have both big and small pores.

There is no satisfactory correlation between the frost resistance of porous ceramics and their strength, water absorption, pore size distribution or saturation coefficient.

See also:
Masonry, Frost attack, p. 55

Salt attack

Porous tiles disrupt when salt solutions in the pores crystallise or react with other constituents and the process is accompanied by a volume increase. The appearance of this kind of disruption is very similar to that caused by frost action, but the presence of salt deposits is usually very noticeable. There are various sources of soluble salts:

- soluble sodium, calcium and magnesium sulphates formed in the tiles during the firing process
- soluble sulphates in the substrate
- sulphate-bearing solutions, such as domestic and industrial effluent
- sodium chloride solutions, such as sea water or road de-icing run-off.

See also:
Masonry, Crystallisation of salts, p. 65

Tile adhesive failure

The first aspect to establish when tiles de-bond from the substrate is whether or not there were opposing dimensional changes in the different elements of the tiling system. This kind of system failure is characterised by doming, bulging or ridging of the tile layer (see the discussion of system failure below).

Tile adhesives can fail because an inappropriate type was used for the specific application or as a result of the ingress of aggressive chemicals.

Inappropriate adhesive

Adhesives of specific type and class are required for adverse conditions, such as cold-room walls, floors subjected to heavy traffic, external facades and flexible substrates. The generic classification of tile adhesives given in Table 4.2 is based on chemical composition and physical form. Note that the development of modern adhesives has resulted in a degree of overlapping of the types, depending on the performance characteristics required. Tile adhesives could thus also be classed according to the application conditions, as shown in Table 4.3.

Table 4.2 *Generic classification of tile adhesives*

Adhesive type	Description
Cement based	A mixture of a hydraulic binder and mineral aggregate (fine sand), with or without a small amount of organic additives. Produced as a dry blend that is mixed with water immediately before use
Polymer based	A ready-for-use compound of polymeric binding agents (aqueous emulsions or latex) and mineral fillers
Reactive resin	The adhesive comprises two or more separate organic components, which polymerise and set when they come into contact. The parts may contain mineral fillers and they are only mixed immediately before use

Table 4.3 *Class classification of tile adhesives*

Class	Description
I: general purpose	For installation of ceramic tiles and mosaics in interior areas of buildings, where the face of the tiled surface is subjected only to intermittent exposure to water
II: water resistant	For installation of ceramic tiles and mosaics in applications where the face of the tiled surface is subjected to prolonged exposure to water
III: extreme conditions	For installation of ceramic tiles and mosaics in applications where high differential movements occur or where high adhesive strength is required (e.g. cold-room walls, floors subjected to heavy traffic, tiling of external facades, tiling on flexible substrates)

Attack by aggressive chemicals
Cement-based adhesives are attacked by sulphates (leached from the substrate or adjoining soil), as well as fertilisers, sewage effluent, industrial chemicals, etc. Note that a sulphate-resisting Portland cement will only be effective against sulphates, but not against acids that may be part of industrial and sewage effluent.

An uncommon failure is the complete disintegration of cement-based adhesives by plaster of Paris, which has been inadvertently or mistakenly added to the adhesive during tiling. Initially the strength development will be rapid, but under long-term and enduring damp conditions the aluminate compounds in the Portland cement and the calcium hydroxide reacts with the sulphate. The reaction is expansive and results in the softening and complete destruction of the adhesive.

See also:
Concrete, Chemical attack on concrete, p. 25

Poor workmanship
Tiles may de-bond at the tile–adhesive interface because one or more of the application characteristics of the adhesive have been exceeded, usually by exceeding the pot life or open time. This type of failure is the result of ignorance, poor workmanship or inadequate supervision.

Preventive and remedial actions

Crazing of the glaze surface, lime blowing or disintegration caused by frost action and salt crystallisation are processes that cannot be stopped by any subsequent treatment or action. They are inherent defects of the tiles. The only remedy if the defect becomes unacceptable is to re-tile the area with sound tiles and by using appropriate adhesives.

- Prevention is critical.
- Careful attention should be given to the choice of materials, correct work procedures and good workmanship.
- Be aware that the application of surface sealants or impregnants to porous, unglazed tiles could exacerbate their spalling due to salt action.
- It is of utmost importance that the correct type of tile and adhesive be selected and that all materials and installation procedures should

There are a number of important application characteristics of tile adhesives.

- *Maturing time*: the interval between the time when the cementitious adhesive is mixed and the time it is ready to use.
- *Pot life*: the maximum time interval during which the adhesive can be used after mixing.
- *Open time*: the maximum time interval since application of the adhesive to the substrate during which the tiles must be embedded in the applied adhesive and still meet the tensile adhesion strength requirement.
- *Wetting capability*: the ability of a combed adhesive layer to wet the tile.
- *Slip*: the downward movement of a tile applied to a combed adhesive layer on a vertical surface.
- *Adjustability*: the maximum time interval during which the position of the placed tile can be adjusted without a significant loss in tensile adhesion strength.

> comply with the relevant national standard specifications and building codes of practice.
- It is strongly recommended that all tiles should be tested for crazing and lime blowing prior to installation.
- If freezing conditions are regularly experienced, tiles or bricks specifically tested for frost resistance, or with a proven frost-resistance record, must be used.

System failure of tiled floors and walls

A tiled floor or wall is a complex multi-component layered system. It consists of various elements:

- The visible parts, namely the tiles, grouting and movement joints.
- The invisible parts, which could include some or all of the following: adhesive, screed or plaster, membrane and slab or wall.

Each element makes a specific functional contribution to the unit as a whole. All the elements have different properties and frequently show incompatible behaviour. As far as the failure of a tiling system is concerned, the dimensional stability of all the components is the determining factor. If all the components of an integrated floor change dimensionally by exactly the same amount in the same direction, then the floor as a unit will expand without any internal stresses being developed.

However, if some elements change dimensionally more than others or in opposite dimensions, internal stresses will develop in the unit. When these internal stresses exceed a critical value, failure of the unit will occur, either by crushing of the tiles or de-bonding of the tiles from the substrate. It is the differential movement which is the determining factor, not the absolute dimensional change of the individual elements.

Differential movement

The reasons for dimensional change of the tiling elements are temperature changes, irreversible moisture expansion of the ceramic tiles, as well as drying shrinkage and creep of the floor or wall.

Thermal movement

The variation in thermal expansion of clay bricks and tiles is small and changes in ambient temperature do not cause significant stresses in a tiled brick wall system.

The difference in thermal expansions of cement products and clay tiles can be significant (see Table 4.4) and, coupled with the diurnal temperature gradient, could be enough to cause stress damage to such tiled systems (e.g. tiled patio floors). Special precautions are required in particular with the tiling of cold-storage spaces.

Irreversible moisture expansion of tiles

The moisture expansion of some ceramic tiles can be considerable. Physically restraining the tiles will not prevent the tiles from expanding. If the floor system is unable to accommodate the stresses caused by the strain, the tiles will either crush or the adhesive will fail and the tiles will lift from the floor or detach from the wall.

Some tiles have an expansion as high as 0.1% (1.0 mm/m) after 5 years of exposure, which is of the same magnitude as their crushing or critical strain. Table 4.5 lists the categories in which tiles are placed according to their total expected moisture expansion (as measured after 96 hours of steaming).

Table 4.4 *Thermal expansion of materials in tiling systems*

Material	Thermal expansion (mm/m per 20°C)
Dense concrete	0.22–0.26
Calcium silicate bricks	0.16–0.28
Fired clay bricks and tiles	0.14–0.18

Table 4.5 *Moisture expansion categories for tiles*

Category	Moisture expansion (%)
Low	0.00–0.05
Medium	0.05–0.10
High	0.10–0.20
Very high	>0.20

Dense, hard-fired tiles tend to have lower moisture expansions than do porous tiles, but aesthetics may dictate the use of low-fired tiles. As far as the tile manufacturer is concerned, the adjustment of the composition of the raw materials is the best approach to minimising the problem of irreversible moisture expansion of the tiles.

Drying shrinkage and creep of cement substrate
The dimensional change of an unsupported concrete slab during the short, medium and long term is very complex. Nevertheless, the combined effect of both drying shrinkage and creep could result in long-term 'shrinkage' at the top surface of a slab by as much as 0.2% (or 2 mm/m).

See also:
Concrete, Dimensional change, p. 29
Masonry, Differential movement, p. 53

Failure of tiled systems

As far as tiled floors are concerned, the general situation is that the tiles want to expand, while the concrete base wants to shrink. If these two elements are not bonded to each other and walls or projections do not restrain the relative movement of the tiles, the tiles will just slide over the base to accommodate the dimensional difference, and thus stresses will not develop in the floor. If the tiled area is restrained around its perimeter, the unbonded tile layer will dome, bulge or ridge.

If all the tiles are bonded to the base, the tiles will be placed in compression, while the base will go into a state of tension. This situation causes the adhesive layer between the tiles and base to be placed in a shear condition. The ability of the adhesive to cope with the shear condition depends on its resilience. If shear failure occurs, the tiles will de-bond and the tile layer will dome, bulge or ridge.

It is of utmost importance to understand that it does not matter whether the expansion of the tiles is bigger or smaller than the shrinkage

The larger the difference between the potential expansion of the tiles and the potential shrinkage of the concrete base, the larger the stresses will be that develop. The compressive force in the tiles is more or less equal to the tensile force in the slab. However, the tiles are much thinner than the slab, and therefore the compressive stress on the tiles is much larger than the tensile stress generated in the slab. It is when the compressive and shear stresses exceed the strength of the components that failure of the floor occurs. The same conditions apply to tiled walls.

of the slab. What matters is that the difference between the expansion and shrinkage does not exceed a critical value. Moisture expansion of the tiles is frequently blamed alone for tiling failures, ignoring the major and sometimes overriding contribution of the dimensional change of the slab or other base materials and adhesive. The fact is that failures can occur even with tiles showing very little irreversible moisture expansion, if there is a large shrinkage of the substrate.

Preventive measures

Failure of tiling systems generally requires retiling of the area. Prevention is thus more important than ever to avoid unnecessary repair costs.

Differential movement

It is very important to realise that moisture expansion is a normal property of tiles and it cannot be curtailed, although the magnitude of the expansion can be minimised by appropriate manufacturing procedures. Similarly, drying shrinkage and creep of cement and concrete is normal behaviour, just as is the settling movements of a structure. It is only when these changes exceed acceptable levels that they are considered abnormal and that the components are judged defective.

If tiles with a critically high moisture expansion must be used (e.g. for aesthetic reasons), the tiles could be given an atmospheric steam treatment prior to installation. The required duration of the treatment to remove enough of the potential moisture expansion should be determined by experiment.

The contribution to tiling failures by drying shrinkages and settling movements of the substrates should not be underestimated. Sufficient time between the wet construction stage and tiling must be allowed for to minimise such movements.

Movement joints

It is a fallacy to think that movement joints will prevent tiling failures due to differential movements between the tiles and the base. These joints are there to accommodate shrinkage, settling and dynamic movements between different parts of the structure. They cannot be used as a cure for poor design, or to avoid the effects of unbridled 'fast tracking'.

Workmanship

It is of utmost importance that the correct type of tile and adhesive be selected and that all materials and installation procedures should comply with the relevant national standard specifications and building codes of practice. In this respect it is necessary to stress in particular the importance of good workmanship. For example, the main reason for detachment of tiles or punching failure on impact is incomplete bedding of the tiles, a condition that is totally avoidable through good workmanship and supervision.

Further reading

Grimshaw R.W. 1971. *Chemistry and physics of clays*. Ernest Benn, London.
National standard specifications and codes of practice.
Rado P. 1988. *Introduction to the technology of pottery*. Pergamon Press, Oxford.
Ryan W. and Radford C. 1987. *Whitewares: production, testing and quality control*. Pergamon Press, Oxford.

Chapter 5

Deterioration of timber

Contents

Agents of attack **88**
Fungi (decay) 88
   Conditions necessary for development of fungi 88
   Wood destroying fungi (rot or decay) 89
   Wood-disfiguring fungi 89
Insects 94
   Beetles (Coleoptera) 94
   Termites (Isoptera) 94
Marine borers 94
Bacteria 96
Weathering 96
Chemicals 98
Heat 98
Mechanical forces 98
Remedial actions **99**
Biological attack 99
   Decay 100
   Insect attack 100
Mechanical and fire damage 100
Weathered timber 100
Preventive actions **101**
Selection of timber 102
   Resistance to decay 102
   Resistance to insect attack 102
Chemical preservatives 102
Timber preservation processes 104
   Preparation 104
   Precautions 104
Providing barriers 104
Environmental control 104
Further reading **104**

Construction timber may deteriorate in service as a result of attack by *biological agents* (fungi, insects, marine borers and bacteria) and/or exposure to *physicochemical agents* (weathering, chemicals, mechanical forces and heat). These factors can, either individually or in combination, cause structural degradation to timber. It is important to note that the occurrence of organisms is restricted to certain geographical locations.

Water plays a very dominant role in the deterioration of timber. Its presence in timber directly affects the properties of wood, such as strength, dimensional change and combustibility, but together with heat its presence also creates conditions favourable for attack by biological agents. Therefore, a wood moisture meter is as essential to the wood scientist and technologist as a stethoscope is to a medical doctor. Furthermore, the first step in any remedial action is to correct the conditions that allowed excessive water to penetrate into the timber.

Although remedial actions may be successful in some cases, such actions generally require specialist services. As always, prevention is better than cure.

Agents of attack

The identification of the type of biological attack, according to the environment, appearance and timber attacked, is summarised in Table 5.1. The names of some economically important fungal or insect species are included to serve as typical examples only. To remain concise, other types of woodborers such as *Bostrychid* spp., wasps and carpenter bees have not been included. The occurrence of organisms depends on specific environmental factors, such as geographical location, climate and micro-environment, as well as on the wood properties.

If proper verification or identification of wood-destroying organisms is required (e.g. for legal documentation purposes) it must be performed by specialist biologists such as mycologists (for fungi) or entomologists (for insects), or, if acceptable, appropriately qualified and registered remedial-treatment technologists.

Fungi (decay)

Conditions necessary for development of fungi
- *Moisture*: wood moisture content 30–50% for optimum development (in certain cases as low as 22% and as high as 60%), depending on type of fungus.
- *Heat*: temperatures between about 20°C and 40°C, sometimes higher (up to 50°C).
- *Other*: adequate oxygen, nutrients and time; the absence of naturally occurring toxic wood extractives or chemical preservatives.

Life cycle of fungi

Fungi germinate from minute *spores*, sending out filaments (*hyphae*) that can form an intertwined mass (*mycelium*). This development can be very rapid. The hyphae penetrate and feed on wood by breaking it down into simple chemical compounds. Under suitable conditions fungi mature to form fruit bodies (*sporophores*), which vary in size from microscopic to large fleshy bodies. Fruit bodies produce vast quantities of spores, which can be spread by air currents, water, birds and animals.

Fruit bodies, strands, mycelium and/or the appearance of the affected wood are used to identify the type of fungus.

Wood-destroying fungi (rot or decay)
- There are many species of fungi.
- Attack can start in unseasoned (green) wood or in previously seasoned, untreated timber. The method of attack of the various destructive species differs and, therefore, remedial treatment also differs.
- Wood-destroying fungi normally do not occur in interior timber unless there is poor design, construction faults or lack of maintenance that cause continuously high moisture contents above 20%.
- The appearance and character of the wood-destroying fungi are summarised in Table 5.2. There are three main types of wood-destroying fungi, if categorised according to the decay found in buildings – i.e. dry rot, wet rot and soft rot. The colour of the decayed wood (i.e. brown rot or white rot) or texture (i.e. soft rot) is also used in the classification of fungi.
- Early stages of decay are virtually impossible to detect. A brown rot may reduce mechanical properties in excess of 10% before a measurable mass loss or visible signs are observed. With only 5–10% mass loss, the mechanical properties are reduced from 20% to 80%.

Wood-disfiguring fungi
- The fungi affect adversely the appearance of the timber, normally without affecting its strength.
- The fungi may create conditions that promote subsequent infestation by decay fungi (rots).
- The appearance and character of staining fungi and moulds are summarised in Table 5.3.

Table 5.1 *Identification of biological attack*

Environment	Appearance	Timber attacked	Cause
Damp, warm interior and exterior conditions Wood moisture content 22–60% Temperature 20–40°C	Attacked timber is dark brown with deep cracks along and across the grain (cubed), and is soft and low in strength. Mycelium appears as white sheets or cotton-like white cushions. Strong musty smell	Mainly softwoods inside buildings	Fungal attack: dry rot (e.g. *Serpula lacrymans*)
Adequate oxygen, nutrients, time Absence of toxic substances such as natural extractives or chemical preservatives	Attacked timber is light or dark in colour, fibrous, not cubed or less cubed than dry rot, very soft and low in strength. Affected timber may form pockets of decay in between sound regions	Hardwoods and softwoods	Fungal attack: wet rot (e.g. *Coniophora puteana*, *Phellinus contiguus*)
	Sections of timber are stained blue. Strength not affected	Sapwood of hardwoods and softwoods	Attack by staining fungi (e.g. sapstain or blue stain)
	Surface of timber covered by powdery or woolly mycelium and spore bodies. Colours range from black, brown and green to orange. Strength not affected	Hardwoods and softwoods	Fungal attack: mould (e.g. *Penicillium* spp., *Aspergillus* spp., *Cladosporium* spp.)
	Round flight holes (c. 2 mm in diameter) and interior of affected timber is riddled with small tunnels. Presence of coarse granular powder of digested wood. Insect is dark-brown and 3–5 mm long	Structural timbers and joinery, both softwood and hardwood. Timber floors are very vulnerable if the subfloor ventilation is poor	Beetle attack: Anobiid borers (e.g. *Anobium punctatum* (common furniture beetle), *Xestobium rufovillosum* (death watch beetle))

Contd

Conditions	Signs	Wood affected	Cause/agent
	Holes on surface and cavities with smooth surfaces filled with granular frass	All types and parts of dry (below the fibre saturation point) timber	Drywood termites (e.g. *Cryptotermes brevis* (powder post termite))
	Tunnel surfaces not as smooth as with drywood termite attack	Damp, decayed or dry wood	Dampwood termites (e.g. *Zootermopsis* spp.)
Wet warm conditions — Wood continuously saturated with water; Temperature 20–40°C; Soil contact; Adequate oxygen, nutrients, time	Surface layer of the wet wood is soft and can be scraped off easily. When the wood dries, the surface develops numerous fine cracks, both with and across the grain. Generally dark in appearance. Microscopic examination is necessary to confirm the cause	All types of wood, but hardwoods are naturally more susceptible to attack	Fungal attack: soft rot (more than 600 species)
Dry warm conditions — Normal humidity; Good ventilation	Round flight holes (c. 1.6 mm in diameter) and the interior of the affected timber is riddled with tunnels. Flour-like powder of digested wood. Insect is reddish brown and 4–5 mm long	Sapwood of certain hardwoods (large pores and high starch content)	Beetle attack: Lyctid borers (e.g. *Lyctus brunneus* (powder post beetle))
	Holes on surface and cavities with smooth surfaces filled with granular frass	All types and parts of wood	Drywood termites (e.g. *Cryptotermes brevis* (powder post termites))

Table 5.1 *Contd*

Environment	Appearance	Timber attacked	Cause
Dry warm conditions – contd	Large oval-shaped flight holes (major diameter 3–10 mm) spaced far apart. Timber surface may bulge due to tunnels near surface. Digested wood consists of fragments and small pellets. Larva (grub) ≤30 mm long. Insect is black and 10–20 mm long	Mainly sapwood of seasoned softwoods, particularly roof timbers	Beetle attack: Cerambycid borers (e.g. *Hylotrupes bajulus* (European house borer))
Timber submerged in sea or estuarine water	Timber riddled with long, round tunnels with smooth shell-like surfaces. Entry holes are very small. Infestation and destruction are rapid	All untreated timber	Molluscs: ship worm; most destructive species is *Teredo navalis*
Timber in marine intertidal zones	Damage appears similar to that by woodborers. Destruction of wooden piles between high and low water marks results in hourglass-shaped timber	All untreated timber	Crustacea: *Limnoria* (gribbles)
Timber in contact with soil	Interior of timber (particularly early wood) is consumed, while the surface is left intact. Attacked timber may be connected to the soil by enclosed runways or galleries made of soil. The insect is large and soft bodied. Destruction is very rapid	All untreated timber. Occurrence is restricted to certain geographical regions	Subterranean termites

Table 5.2 *Wood-destroying fungi*

Fungi	Appearance and character
Dry rots	Typically found in timber in direct contact with wet concrete or brickwork; also in inefficiently ventilated areas (e.g. subfloor spaces)
	Rarely found in timber exposed to fluctuating conditions
	Some dry rots are also called brown rot. They feed mainly on cellulose, leaving darker (brown) lignin-rich residue
	Attacked timber typically has deep cracks along and across the grain, is soft and low in strength
	Mycelium appears as silky white sheets or cotton-like white cushions
	Dry rots attack mainly softwood
	Serpula lacrymans (commonly called the dry rot fungus) is a major decay fungus
Wet rots	Many species exist. They feed on both cellulose and lignin
	Coniophora puteana (a brown rot called cellar fungus) is found in water-soaked hardwood or softwood. Cross-cracks may be not as deep as those produced by *S. lacrymans*. Freshly colonised wood appears yellow
	Poria species, such as *Antrodia vaillantii* (white pore or mine fungus), generally attack softwoods and timber in locations of higher temperature. Infected wood breaks into cubic pieces. Can be confused with *S. lacrymans* affected wood
	Phellinus contiguus, a white rot fungus, is commonly found in exterior hardwood or softwood joinery. Attacked wood is light coloured, has a stringy, fibrous texture, is not cubed and does not crumble
	Asterstroma spp., a white rot fungus, is usually found in softwood joinery such as skirting. The wood texture is similar to that of *P. contiguus* infected wood
Soft rots	Regarded as superficial forms of wet rot. Develops in very wet conditions (e.g. timber in ground contact). The chemical character of attack is different from that of brown or white rot
	Fungi make longitudinal cell wall cavities, leaving the surface layer soft
	Soft rots attack all types of timber

Table 5.3 *Wood-disfiguring fungi*

Fungi	Appearance and character
Staining fungi (blue stain in service)	Attack is visible as dark-blue streaks and patches in the timber or underneath transparent or semi-transparent surface coatings on timber
	Feed on the cell contents (sugars and starches in the ray cells of sapwood), not on the cell wall (cellulose, polyoses, lignin). Penetrate sapwood
	Most common type is *Aureobasidium pullulans*
Moulds	Growth is most severe on sapwood
	Fungi produce powdery or woolly mycelium and masses of spores at the surface
	Colours range from black, brown and green to orange
	Can also grow on timber treated with preservatives
	Many types exist, commonly *Penicillium* spp., *Aspergillus* spp. and *Cladosporium* spp.

Note

Do not confuse the localised green or dark discoloration that develops as a result of chromated copper arsenate (CCA) or other copper containing treatments of timber with the discoloration caused by blue stain fungi. This discoloration is particularly noticeable in sapwood regions of softwood that has been exposed to direct sunlight soon after treatment.

Insects

Beetles (Coleoptera)
The most important beetles are contrasted in Table 5.4.

Termites (Isoptera)
The character and occurrence of the three types of wood-destroying termites are summarised in Table 5.5.

Marine borers

- Found along all coastlines, most active in warm waters and estuaries.

Table 5.4 *Wood-destroying beetles*

Borer	Infestation	Appearance
Lyctid borers (e.g. *Lyctus brunneus* (powder post beetle))	Infest only the sapwood of hardwoods with large pores and enough starch Hardwoods with a high sapwood content (e.g. *Eucalyptus* spp., kiaat, oak) are most in danger	Small, round flight holes (1.6 mm) and fine, flour-like frass
Anobiid borers (e.g. *Anobium punctatum* (common furniture beetle), *Xestobium rufovillosum* (death watch beetle))	Attack hardwoods and softwoods. Typically attack old furniture Need damp, humid conditions, and thus timber floors are vulnerable if the subfloor ventilation is poor Attacks are serious once started, and do not stop of own accord	Round flight holes (2 mm) and coarse granular powder of digested wood, small heap of frass around flight holes in horizontal surfaces
Cerambycid borers (e.g. *Hylotrupes bajulus* (European house or longhorn borer), many other species)	Very serious threat to construction timber, but restricted to particular geographical regions Attack mainly sapwood of softwoods	Normally invisible until oval-shaped flight holes (3–10 mm) appear. Frass marks on irregular surfaces

> There are four distinct stages in the insect life cycle: egg, larva, pupa and adult beetle. The larval period varies from 1 to 10 years depending on the species and environmental conditions.
>
> The larva ('woodworm') burrowing into wood for food does most damage. After pupation the insect turns into an adult inside the wood, close to the surface, and cuts flight (emergence) holes through the surface of wood. Wood powder (or frass) is often visible around the holes.

• All types of timber are susceptible to attack.

The infestation and appearance of timber attacked by molluscs and crustaceans are summarised in Table 5.6.

Table 5.5 *Wood-destroying termites*

Termite type	Character	Occurrence
Subterranean termites (e.g. *Coptotermes* spp., *Mastotermes* spp., *Microcerotermes* spp.)	Soft bodied, cannot survive in the open, depend on a source of moisture, thus live underground in nests and forage in enclosed galleries Eat the interior of timber (particularly early-wood), leaving the surface intact Entry holes and breakthroughs are plugged with soil	Attack wood in contact with soil. Short connecting galleries may be constructed
Dampwood termites (e.g. *Zootermopsis* (north-west USA and west Canada))	Dependent on a source of moisture (i.e. wood in contact with soil) Narrower tunnels in dry than in damp wood	Initially infest damp, decayed timber and later dry, undecayed timber (e.g. buildings and bridges)
Drywood termites (e.g. *Cryptotermes brevis* (powder post termite))	Do not require a source of moisture; live in wood out of ground contact Greatly influenced by the air temperature and relative humidity Faecal pellets are poppy-seed shaped. Tunnel in all directions	Infest dry sound timber (e.g. inside buildings located in coastal regions)

Bacteria

- Bacteria are early colonisers of wood exposed to soil and aquatic (sea or fresh water) environments or to out of ground contact conditions where wood is subjected to periodic wetting and drying.
- Attack cell walls by cavitation, erosion and tunnelling, causing structural damage.

Weathering

- Exposed timber deteriorates due to the combined effect of sunlight, heat and water.
- The UV component of sunlight mainly breaks down lignin (a

Termites are social insects and live in large colonies, with a wide distribution in the tropics and certain temperate regions. Although frequently referred to as 'white ants', there is only a superficial resemblance to true ants. Three types of termites exist: subterranean, dampwood and drywood termites. Colonies consist of workers, guards, males and females, and, centrally, a queen capable of lying up to 80 000 eggs per day. Because these soft-bodied insects cannot survive in the open, large nests are constructed from cemented soil, either underground or above ground depending on the species. New colonies are formed by the characteristic swarming of large numbers of winged individuals.

The food of some termites is grass, leaves and wood in a state of incipient decay. Several species are a significant risk to any wood product in buildings, such as structural timber, furniture and even books. The interior of the articles may be totally consumed, leaving an apparently intact surface.

Table 5.6 *Marine borers*

Marine borer	Infestation	Appearance
Molluscs (shipworm); most destructive species is *Teredo navalis*	Infestation and destruction of submerged timber can be very fast Holes of up to 10 mm in diameter opening to the exterior	Very long round tunnels with smooth shell-like surfaces that riddle the infested timber
Crustaceans (*Limnoria* (gribbles))	Occurs in temperate waters Less prevalent and virulent than shipworm; may take years to destroy submerged timber	Attack is from the surface, with pits and holes of 2 mm or less in diameter appearing only after many months or years Damage is similar to that done by wood beetles, in wooden piles mostly between the high- and low-water marks (hourglass-shaped piles)

structural component of wood), resulting in a bleached and, later, grey appearance of the timber. The penetration of the radiation is low and its effect is thus limited to the surface layers. The process is very slow and mainly affects the aesthetic appearance of the timber.
- Dust-laden wind abrades the surface layer. The softer early-wood is eroded preferentially, resulting in a characteristic roughened surface appearance.
- Cyclic wetting by rain, dew and other forms of precipitation and subsequent drying cause swelling and shrinkage. Continued repetition of this dimensional change causes surface checks to form. Serious weathering can lead to warping, cupping and, later, cracking. Apart from being unattractive, this process opens the way for fungal spores to penetrate the wood and ultimately contribute to structural failure.

Chemicals

- Wood is naturally mildly acidic (pH 2 to 5.5) and thus resistant to weak acids. Strong acids and alkalis will cause damage.
- Affected timber shows bundles of soft cotton-wool-like fibres that are attached to the surface. The fibres are visible in, for example, softwood roof timber underneath fibre cement or ceramic roof tiles.
- Rusting of steel nails in wet timber is a common cause of chemical attack. The metallic corrosion process creates hydroxyl ions, resulting in highly alkaline conditions around the nail, which break down the structural components of the wood. The attack around steel nails and fittings is accelerated if salt-type preservatives are used. Salt-laden sea air will also accelerate the attack.
- Steel fasteners can also react with wood extractives (e.g. tannins) causing dark blue stains.

Heat

- The immense heat generated by combustible material indoors is primarily responsible for the fire damage to timber in constructions. Large wooden members require a constant source of heat to start burning.
- Charring rates for structural timber range from 15 to 20 mm in 30 minutes.

Mechanical forces

- Unbalanced tensional, compressive, bending and torsional forces can cause permanent deformation by creep, or even structural damage such as splits, cracks and breaks along and across the grain. Normally this kind of failure is prevented by proper design and building practices.

- Localised pedestrian or vehicle traffic over wooden surfaces such as steps, decks, bridges and thresholds, leads to abrasive damage of wooden surfaces.
- Swelling and shrinkage caused by changes in moisture content could result in mechanical damage.

Remedial actions

Biological attack

The most typical remedial actions against biological attack are summarised in Table 5.7.

Table 5.7 *Remedial actions against biological attack*

Attacking agent	Remedial actions
Insects	
Wood borers (beetles), drywood and dampwood termites	Fumigation, drill and inject, spraying, flooding with preservative solution, replacement with treated wood where necessary, etc.
Subterranean and dampwood termites	Soil poisoning, baiting, physical barriers around treated posts, replacement with treated wood where necessary, etc.
Fungi and bacteria	
Wood-destroying fungi and bacteria	Eliminate source of moisture, replace infested wood, insert fused rods containing preservatives, pole bandages at ground line, etc.
Disfiguring fungi	Moulds: biocide surface treatment
	Sapstain: replacement of wood or application of opaque surface coating
Marine borers	Replacement of affected timber with appropriately treated timber

Important: The use of biocides is strictly controlled at national and regional levels. The relevant regulations must be consulted and adhered to.

Decay
- The identification and extent of the type of decay (i.e. dry rot or wet rot) and the subsequent selection, planning and execution of appropriate remedial action is a specialist task.
- The source of moisture must be established and eliminated. If ingress of moisture cannot be avoided, all timber should be replaced with appropriately pretreated wood. Such treatment should, if possible, also fulfil a waterproofing function.
- Adequate ventilation should be established.
- Appropriate heat treatment (ambient temperature raised to about 50°C) can be considered.
- The attack by some species of fungi is very insidious and extensive. In such cases, severe surgery and even replacement of all timber may be required.
- Chemical preservation is accomplished with a variety of fungicides applied in organic solvents, fused rods, pastes or with bandages.

Insect attack
- The identification and extent of the type of insect attack and the subsequent selection, planning and execution of appropriate remedial action is a specialist task.
- Treatment with an insecticide preservative may be effective, if the attack has been detected during a very early stage.
- Fumigation with, for example, methyl bromide is very effective at stopping attack. This treatment leaves no residual toxic substances after treatment.
- Light organic solvent preservatives are widely used in this role because they can be applied easily *in situ* to the affected timber by drill and injection, or by flooding, spraying or brushing. The timber also remains dimensionally stable.
- Proper precautions must be taken during the application of the preservatives, as the vapours of the solvent are flammable and toxic.
- If the attack is severe, replacement of the affected timbers will be required.
- Appropriate heat treatment can be considered.
- Termiticides to eradicate subterranean termites are applied by drilling into foundations.

Mechanical and fire damage
Where such damage occurs, timber is normally replaced.

Weathered timber
- Damaged surface coatings are chemically (paint stripper), thermally (heat gun) or mechanically (sanded) removed.

- After proper surface preparation, such as bleaching, filling of cracks and holes, and sanding, the coating system is applied according to the recommended instructions.

Preventive actions

Preventive actions to minimise the deterioration of timber are summarised in Table 5.8.

Table 5.8 *Preventive actions to minimise deterioration of timber*

Action	Comments
Timber selection	Only use timber with the required durability (natural or treated)
Naturally durable	Heartwood sections only
Preserved timber	Treated according to relevant (national) specifications. The type, penetration depth and retention of preservative are specified for each application environment and wood product
Provide barriers	
Marine borers	Sheath submerged marine timbers with copper plates, concrete collars, etc., where appropriate
Termites	Treat soil around buildings with insecticides; install physical barriers around treated foundation poles
Fire	Impregnation with fire-retardant chemicals or application of surface coatings
Weathering	Protect exposed timber with well-maintained surface coatings such as oils, varnishes or paints. Follow the manufacturer's instructions and recommendations
Environmental control	
Fungi	Ensure interior spaces remain well ventilated. Wood rot cannot develop or exist unless the moisture content of the timber is above about 20%. The moisture content of timber exposed in dry interior spaces is generally well below 18%
Fire	Install smoke detectors and sprinkler systems

Selection of timber

Resistance to decay

- Durability classifications of timber relate to the resistance against fungal attack of the heartwood of the species. The sapwood is regarded as non-durable.
- There is wide variation in the natural durability of timber, with some hardwoods (e.g. teak, jarrah and iroko) classified as very durable. However, many hardwoods are non-durable or even perishable (e.g. meranti, some *Eucalyptus* spp., elm, American red oak, birch and willow), while certain softwoods are classified as durable (e.g. yew, sequoia and red cedar). Specifications for durability should thus be species specific.
- The sapwood of both hardwoods and softwoods are susceptible to decay. Most construction timber has a high proportion of sapwood, and treatment with appropriate preservatives is thus required.

Resistance to insect attack

- Woodborers mainly attack the sapwood. In the case of *Lyctid beetles*, the female inserts her ovipositor into the micro-pores of the sapwood. The micro-pores of certain hardwoods are too small to accommodate the ovipositor and these timber species are thus immune from attack by this type of woodborer. Female *Anobiid borers*, on the other hand, lay eggs on the rough surface of end grains or in small cracks and crevices, and thus the size of the micro-pores of the timber is irrelevant.
- Attack by woodborers is restricted to certain geographical and climatic regions. Regional or government regulations stipulate districts within which the appropriate treatment of timber for structural use is prescribed.

Chemical preservatives

The classification and types of chemicals used in the preservation of timber are summarised in Table 5.9. Some compounds or formulations are banned in some countries or their continued use may be under review.

Important

The use of insecticides and preservatives is strictly controlled by national and regional regulations. The relevant regulations must be consulted and adhered to.

Table 5.9 *Classification of chemical preservatives*

Preservative type	Description
Water-borne	Treatment wets timber, and thus grain raising and deformation (caused by differential shrinkage) may occur
	Odourless and clean. Timber can be painted, stained or glued when dry
	Borate compounds:
	– deep and fast penetration by diffusion
	– soluble, leaching must be prevented
	– colourless, only for interior or well-protected exterior applications
	– a properly maintained coating is recommended
	Copper–chrome–arsenic (CCA) and similar formulations:
	– copper acts as a fungicide and arsenic as an insecticide; chrome reacts in the presence of wood cellulose to chemically fix the copper and arsenic as insoluble compounds in the wood
	– stains wood light green
	– corrosive; use only hot-dipped galvanised or stainless steel fasteners
Oil-borne	*Creosote (a coal distillate)*:
	– heavy-duty fungicide and insecticide
	– preservation process does not cause dimensional change of the treated timber
	– highly water repellent and leach resistant
	– strong odour, exterior use only; cannot be painted or glued readily
Preservatives dissolved in light organic solvents	Non-swelling, non-corrosive, non-staining
	Solvents are highly flammable and toxic, and are a serious health hazard until evaporated
	Many compounds: chloronaphthalenes, lindane, organotin compounds (e.g. TBTO), pyrethroids, organoboron compounds, pentachlorophenol (PCP), etc.

Timber preservation processes

- The choice of preservative and method of application is dictated by the severity of the biological hazard, exposure conditions, wood properties and the required degree of penetration and retention of the preservative.
- A variety of processes are available (e.g. diffusion, brushing, spraying, immersion, pressure, vacuum/pressure) and different combinations of processes may be used.

Preparation
- The timber must be correctly seasoned before preservation. Except for waterborne diffusion treatment processes, the moisture content must be below the fibre saturation point (25–30%).
- All machining should, as far as possible, be carried out before treatment.

Precautions
- The correct and safe use of preservatives in accordance with label instructions and legal requirements should not normally endanger the health of people or non-target animals.
- Always use dust masks and goggles when machining treated timber.
- Wash hands and face before eating.
- Preservatives are toxic to fungi and insects, but the use of treated wood should, under normal conditions, not be a health hazard to humans and animals. Care is needed where birds, fish and bees may come into contact with treated wood.
- *Never* use treated wood (particularly copper–chrome–arsenic (CCA) treated timber) as firewood.

Providing barriers

Specific techniques to control attack by marine borers, termites, fire and weathering are summarised in Table 5.8.

Environmental control

Specific environmental control measures to reduce possible damage by fungi and fire are summarised in Table 5.8.

Further reading

Building Research Establishment 1972. *Recognising wood rot and insect damage in buildings*. Report. Department of the Environment, Building

Research Establishment, Princes Risborough Laboratory, Princes Risborough.

Eaton R.A. and Hale M.D.C. 1993. *Wood. Decay, pests and protection.* Chapman & Hall, London.

Forest Products Laboratory, US Department of Agriculture 1989. *Handbook of wood and wood-based materials for engineers, architects and builders.* Hemisphere, Madison, WI.

National standard specifications and codes of practice.

Richardson B. 1978. *Wood preservation.* Construction Press, Lancaster.

Wilkinson J.G. 1979. *Industrial timber preservation.* Associated Business Press, London.

Entry web sites

American Wood-Preservers' Association, USA: http://www.awpa.com

British Wood Preserving and Damp-proofing Association, UK: http://www.bwpda.co.uk

Building Research Establishment, UK: http://ww.bre.co.uk

Federal Research Centre for Forestry and Forest Products, Germany: http://www.bfafh.de

Forest Products Laboratory, USA: http://www.fpl.fs.fed.us

Forintek Canada Corporation and the Canadian Wood Council: http://www.durable-wood.com

International Research on Wood Preservation, Sweden: http://www.irgwp.com

Chapter 6

Paint failures

Contents

Paint failures **107**
Causes of failure **108**
Instability of the substrate 108
Improper preparation of the substrate 108
    Metals 109
    Concrete and other cement products 112
    Timber 112
Using unsuitable paints 113
Moisture in substrates 113
Quality of paint 114
Preventive and remedial actions **114**
Further reading **115**

Building structures present a variety of substrates that may require painting. The most common substrates are concrete, plaster and rendering, steel and other metals, timber and wood products. These vary widely in physical and chemical properties and some are difficult to paint satisfactorily. The paint or coating is normally required to perform a decorative or protective function. Paint technology is complex and will not be considered, this chapter being concerned only with paint failures that may be observed in the normal course of building defect investigations.

Paint failures

Whether stable or unstable, all surfaces over which paint is applied will eventually require re-coating, either for protective or for decorative reasons. Paint coatings are not permanent, and when one considers that they are basically organic in nature and that the two or three layers applied during the application of a paint system yield a total dry film thickness of only 50–100 μm, one ought not to be surprised that they require renewal at some stage. A breakdown of a paint coating is, therefore, not necessarily a fault unless it occurs within a short time after application.

The life of a paint coating is dependent on many factors, the most important being:

- the type of substrate
- the preparation of the substrate before paint application
- the formulation of the coating
- the application procedure and the climatic and service conditions.

Some coatings have failed within a few days of application, while others have been exposed to the weather, in service, for more than 18 years and are still in a satisfactory condition. Five years seems to be a commonly accepted painting cycle for exterior work. Any serious failure occurring within a shorter period may be considered a fault.

As surface finishes, paints and related coatings are dependent on the overall condition of the substrate and the immediate environment. Therefore, paint failure is often indicative of other serious building problems, such as water penetration, because of defective detailing or related deterioration and extreme exposure of certain building elements. The most widely occurring types of paint deterioration are summarised in Table 6.1.

Causes of failure

Paint failures can be divided into a number of broad classes.

Instability of the substrate

Paint failures are often a result of painting too early on 'green' surfaces or are due to a poor choice of materials or a bad design. No paint can breach a crack that forms in a substrate after it has been painted. Clinker-ash bricks, for instance, can sometimes exhibit lime popping, which throws off paint most effectively. Timber that is wet during painting or that has been treated with preservatives with a high boiling point solvent as a carrier can cause paint blistering.

Improper preparation of the substrate

The most common reason for improper preparation of substrates before painting is poor site supervision. Surfaces may also be weak and friable, contaminated, excessively alkaline, cracked or contain soluble salts. Steps may be taken to mitigate these conditions, but paint applied to a poor substrate cannot produce a durable finish.

Metals
The main cause of early failure is painting on surfaces from which corrosion products or scale have not been adequately removed. If the occurrence of corrosion is extensive or severe, abrasive cleaning or grit blasting is recommended (wire brushing is ineffective to remove heavy rust).

Galvanised steel sheeting and sections or units require special attention. Galvanised steel sheeting often shows 'white rust' or 'storage stains'. Where this occurs, the rust is often not removed before painting and flaking or peeling of the superimposed paint inevitably ensues at an early date. Various waxes and passivation agents are sometimes applied after galvanising in order to minimise white rust formation. The recommended preparation procedures for galvanised surfaces are:

1. Thoroughly clean the surfaces by washing with a solvent or detergent. So-called 'wash primers' are not effective cleaners. In difficult cases abrasive materials may be used judiciously, but care should be taken to avoid excessive removal of the zinc coating.
2. Apply a mordant wash (a solution of copper salts in phosphoric acid and an organic solvent) to heavily galvanised sections only (e.g. those galvanised by hot dipping).
3. Prime with polyvinyl butyral or acrylic resin-based etch primers or zinc phosphate based primers.

Excessive thickness of zinc-rich primers can cause pealing. Calcium plumbate primers formulated for direct application to zinc surfaces are satisfactory, subject to the use of compatible topcoats. These primers may not be generally available, because lead-based paints are regarded to pose a health problem.

Hot-rolled, mill-finished stainless steel can be successfully painted provided that surface preparation and priming is meticulously carried out:

1. *Surface preparation.*
   (i) Remove all residues and loose deposits. Scaled or weathered surfaces should be blast cleaned, but blasting media *must* be free from metallic iron, iron oxides, other metals and chlorides.
   (ii) Thoroughly degrease the surface with a water-dispersible cleaner.
2. *Primer*: twin-pack primers (e.g. polyvinyl butyral based resin with phosphoric acid as the second component) must be used to ensure good adhesion.

Table 6.1 *Common failures in paint systems*

Failure	Description	Causes
Blistering, peeling or flaking	Most common deterioration problems	Improper surface preparation
	Characterised as a general loss of adhesion between layers or at the substrate interface. The paint film remains largely intact	Incompatibility between overlying paint films (e.g. oil and latex emulsions)
		Entrapment and migration of water or water vapour behind the paint system
		Rust formation underneath the paint film
		Rain early in the exposure life of the paint film
Solvent blistering	Dome-shaped projections in paint film	Entrapment of volatile components of the fresh paint as a result of too rapid drying of the paint film
Chalking or powdering	Formation of friable or dusty powder on the surface	Gradual breakdown at the surface of the binder through photochemical degradation, usually from UV radiation
	Degraded material can cause staining to adjacent materials	
Wrinkling	Fine, wave-like pattern, frequently in patches	Paint films applied too thickly
		Paint applied before the previous coats have dried
		Paint containing solvent that dissolves previous coats
		Paint applied to a too cold surface

Checking, crazing, or surface micro-cracking	Small cracks or splits that penetrate the top layers only or extend completely through to the substrate	Poor-quality paint Paint films have become too brittle and are no longer able to expand and contract in response to changes in the ambient temperature and humidity or dimensional changes in the substrate
Discoloration	Black, brown and green surface discoloration	Growth of micro-organisms on the surface Biological attack from micro-organisms may discolour paint
Staining from the substrate	Rust stains running down the painted surface	Oxidation of the metal substrate
Resin eruptions	Viscous resin oozing through the paint film	Bleeding of resinous knots in poorly prepared wood
Photochemical degradation	Light-sensitive pigments fade and discolour	UV degradation, which depends on the composition of the paints and the pH of the environment Oils can darken through oxidation

3. *Finishing coats.*
    (i)   Twin-pack epoxy resins for interior use and a high gloss finish.
    (ii)   Twin-pack polyurethane/acrylic based resins for interior or exterior use and a high gloss finish.
    (iii) Single-pack, waterborne acrylic resins for interior or exterior use and a matt or sheen finish.

Consult paint manufacturers regarding the best generic paint system for the expected environmental conditions.

See also:
  Metal, Structural steel, p. 38
  Metal, Galvanised steel, p. 42
  Metal, Stainless steel, p. 44

Concrete and other cement products
Building materials based on Portland cement, for instance, are generally porous and initially saturated with water carrying calcium hydroxide. When applied to fresh cement surfaces many paints tend to fail through loss of adhesion to the substrate, because of water vapour pressure or the build up of efflorescent salts at the paint–substrate interface. The alkaline nature of the surface also causes disintegration of some paint binders by saponification.

Unless these factors, together with considerations of service conditions, are taken into account when paints are being specified for use on cement-based surfaces, the risk of early failure is great. Normally, suitable finishes may also be unsatisfactory under certain conditions (e.g. gypsum-based plaster under moist conditions).

One of the reasons for the failure of emulsion paints on gypsum skim coats is that the water in the paint softens the water-sensitive gypsum, which re-crystallises and so produces a poor bond. Gypsum skim coats must be treated with an oleo-resinous (bonding) liquid before painting with emulsion paint.

See also:
  Masonry, Cement-based mortar and plaster, p. 60

Timber
A change in the moisture content of timber causes considerable and anisotropic dimensional changes. Timber should thus not be painted before equilibrium with the environmental humidity has been reached, otherwise the paint will crack and de-bond. The cross-cut ends of timber should always receive special attention, because this is the most vulnerable surface for ingress of moisture. Achieving good adhesion with dense or oily wood is difficult, and specialist advice should be sought.

See also:
   Timber, Weathered timber, p. 100

Using unsuitable paints

Choosing unsuitable paint systems for a particular job is a mistake that is all too easy to make. The variety of raw materials used in the paint industry has increased to such an extent that the paint user is faced with a bewildering assortment of available coatings. Generic names given to paint types can be misleading, since some characteristics may be considerably modified in a particular formulation. However, good guidance is given in different national codes of practice on many paint systems. New and improved paint systems are continually being developed and the onus rests heavily on the paint manufacturer to provide reliable guidance regarding choice and usage.

Some paints have acquired a reputation for success. For example, acrylic emulsion paints have proved to be so effective on cementitious substrates that it has become common to specify them for this purpose. However, such a paint applied to a cold, dense, off-shutter concrete may crack and fail.

The characteristics of a paint, such as the tensile strength of an epoxy or the extensibility of a polyurethane resin, although high compared with other coatings, should not lead one to believe that either could be used on a severely hair cracked plaster surface.

The choice of paint should be dictated by the total lifetime cost, not by material cost alone. While different brands may vary in price, the quality of paint is usually directly related to its cost.

Moisture in substrates

The presence of moisture in substrates is the most common source of paint problems. For instance, when exposed to the weather, timber absorbs, retains and releases moisture according to the relative humidity of the atmosphere at different times. During this process shrinking, warping and cracking can take place, depending on the density and other characteristics of the timber. Under such circumstances it is difficult for superimposed coatings to adhere well to the substrate, and flaking may take place.

Excessive moisture in the substrate affects adversely the application of most types of paint as well as their long-term performance. It can cause the deterioration of many materials and promote the growth of micro-organisms.

As a rough estimate it takes 1 month for every 25 mm thickness of wet construction to dry to an acceptable level. Such long periods are usually unacceptable and there is a tendency to apply paint too soon. The surface of a substrate may dry quite quickly and, provided that one avoids a glossy finish (usually relatively impermeable paint), it can be painted with a high-permeability emulsion paint as a temporary solution until the substrate is dry enough to take a permanent finish. Forced ventilation, which is very often neglected, and heat both speed up the drying of interior surfaces.

Even when paint has been applied to an apparently dry surface, moisture may later migrate to the interface, causing blistering, loss of adhesion and eventual breakdown of the paint film. Heavy water shedding on elements of structures also affects detrimentally the performance of paints.

Quality of paint

In some instances the poor quality of the paint is the reason for failure, but it is difficult to separate this from 'the wrong choice of paint'. Adulteration of the paint can be put down to bad painting practice. In some cases this may be plain dishonesty on the part of the applicators, while in others it is due to ignorance.

Preventive and remedial actions

The more common types of painting problems encountered on the building site that cause much time, labour and materials to be spent on repair work can, to a certain extent, be obviated by:

- designing buildings in such a way as to prevent the ingress of rainwater into the walls and other elements of structures
- giving consideration to the painting aspects of the building or structure at the planning stage.

There are various aspects of good practice that must be adhered to:

- Procure reliable advice regarding paint and painting procedures from experienced consultants with a proper knowledge of paint types, an understanding of the physical and chemical properties of the substrates over which the paints are to be applied and of the effects of the pertinent macro- and micro-climatic conditions on both the paints and the substrates.
- Paint systems that require redecorating (an existing paint surface that is in good condition but is slightly faded and worn, or where a new colour scheme is required) must still be thoroughly prepared. Spot patching, light sanding to remove small irregularities or blemishes and washing down (e.g. with sugar soap) to remove dirt and greasiness are important steps.
- Ensure that substrates that will be exposed to the elements and which

require painting are in a good and stable physical condition.

- Properly prepare the substrate. Do not allow unnecessary sharp edges on timber or metal. Blisters on painted metal and paint flaking over wood surfaces invariably start at sharp edges when these are present. Owing to surface tension effects paint tends to run away from edges on drying, leaving these areas poorly protected, which often gives rise to premature failure.
- Insist on adequate site supervision during paint application.

Do not:

- use paint types other than those specified by a recognised authority on paints
- apply paint systems without the specified undercoats
- paint during adverse climatic conditions
- apply successive coats without adequate intermediate curing times
- over-thin the paint and ignore the specified spreading rates
- apply catalysing type paints without the catalyst.

Further reading

Building Research Establishment 1982. *Painting walls*. Digests 197 and 198. Building Research Establishment, Watford.
Building Research Establishment 1998. *Internal painting*. Good Repair Guide 19. Building Research Establishment, Watford.
National standard specifications and codes of practice.
Oil and Colour Chemist's Association, Australia 1983. *Surface coatings*. Tafe Educational Books, Randwick.

Entry web sites

Federation of Societies for Coating Technologies, USA: http://www.coatingstech.org
National Association of Corrosion Engineers, USA: http://www.nace.org
Paint Quality Institute home page: http://www.paintquality.com
Paint Research Association, UK: http://www.pra.org.uk
Painting and Decorating Contractors of America: http//www.pdca.com
Society for Protective Coatings, USA: http://www.sspc.org

Chapter 7

Examples of failure

Contents

Concrete failure **118**
Chloride-induced corrosion of reinforcing steel 118
Carbonation-induced corrosion of reinforcing steel 119
Sulphate attack of a concrete slab 120
Alkali–aggregate reaction of concrete 121
Failed repair of a reinforced concrete beam 122
Metal failure **124**
Corrosion of galvanised reinforcing steel 124
Corrosion of a wrought-iron railing 125
Corrosion of iron at setting points 126
Red rusting of galvanised roof sheeting 127
Deterioration of masonry **128**
Irreversible moisture expansion of brickwork 128
Differential movement in brickwork 129
Lime blowing in concrete blocks 132
Salt action and efflorescence 133
Chloride attack of calcium silicate bricks 135
Sulphate attack of mortar 136
Frost attack 137
Tiling failure **138**
Failure due to exceeding the open time of the adhesive 138
Failure of external tile cladding 139
Tiling failure caused by differential movement 140
Lime blowing in floor tiles 141
Timber deterioration **142**
Wet rot of pergola timbers 142
Wet rot of a hardwood door 143
Wood-disfiguring fungi 144
Insect attack: Cerambycid borers 145
Insect attack: Anobiid borers 146
Powder post termite 147
Marine borers 148
Moisture damage 149

Paint failure **150**
Flaking from new cementitious substrate 150
Delamination of newly painted surfaces 150
Staining by micro-organism growth 151
Paint failure on an exterior hardwood door 152
Redecorating a garden gate 153
Painted galvanised steel 154
Paint failure of galvanised roofing sheets 155

Concrete failure

Chloride-induced corrosion of reinforcing steel

Refer to:
> **Concrete, Corrosion damage to reinforced concrete, p. 2**
> **Concrete, Chloride testing, p. 7**

This failure of a concrete slab was caused by severe corrosion of the reinforcing steel. Although the concrete cover to the steel was adequate,

Photo: *J.E. Krüger, Pretoria, Republic of South Africa*

the steel nevertheless corroded severely because calcium chloride was mixed into the concrete as an accelerator.

Remedial work is far from straightforward and would best be left to a specialist in the field of concrete repair.

Carbonation-induced corrosion of reinforcing steel

Refer to:
Concrete, Manifestations of corrosion damage, p. 2

Photo: *J.R. Mackechnie, Cape Town, Republic of South Africa*

A typical example of carbonation-induced corrosion is illustrated. This particular structure is a foreshore freeway bridge, about 25 years old. The cover of the reinforcement now exposed was originally 10–20 mm.

Carbonation-induced corrosion causes general corrosion of reinforcement with multiple pitting along the steel surface. The corrosion rates are moderate because carbonated concrete tends to have fairly high resistivity that discourages macro-cell formation. Steel exposed to corrosive conditions will therefore show signs of corrosion locally in the form of rust stains or cracking. Usually, carbonation-induced corrosion is limited to low cover depths, typically less than 20 mm. Reinforcement at greater depths is often protected from corrosion, despite being embedded in carbonated concrete, because of the high resistivity of the material.

More severe corrosion of reinforcement may occur when carbonation is associated with the presence of salts in concrete. Chlorides admixed into concrete are not problematic until carbonation releases bound chlorides around the reinforcement, causing intense pitting corrosion of the steel.

Sulphate attack of a concrete slab

Refer to:
Concrete, Chemical attack on concrete, p. 25

Photo: *J.E. Krüger, Pretoria, Republic of South Africa*

This structural failure of the foundation wall of a house in a mining community was caused by the expansion of the concrete floor slab on grade. The concrete was made with mine waste with a very high concentration of sulphides. The slab was cast directly on grade, with no moisture barrier membrane, and thus the concrete remained damp. The inevitable sulphate attack caused a very large expansion of the slab, pushing the foundation walls outwards. The use of aggregate of this kind should be avoided, but appropriate design precautions must be taken, if no other aggregate is available. Repairing the brickwork without design changes will not solve the problem, as the concrete will continue to expand for a long time. Note also the deterioration on the bottom courses of the foundation wall that was caused by sulphate salt crystallisation.

Compare this failure with the example of irreversible moisture expansion of fired clay bricks (see p. 128), which has a similar appearance.

Alkali–aggregate reaction of concrete

Refer to:
Concrete, Recognition of alkali–aggregate reaction, p. 22

Photo: *J.R. Mackechnie, Cape Town, Republic of South Africa. Scale: width 400 mm*

Map cracking represents a low level of distress but, although it is the most obvious indicator of alkali–aggregate reaction (AAR), it is not diagnostic and further testing was required for positive identification. In situations where AAR is well advanced, the crack pattern is much better developed and some silica gel becomes evident at the surface.

Alkaline pore solutions react with metastable phases in certain types of aggregate to form a reactive gel that expands when it imbibes water. The swell pressure causes the concrete to disrupt and expand internally, which is manifested by cracking at the surface. In unreinforced concrete the cracks characteristically form a map pattern. Macroscopic and microscopic examination of drilled cores is always essential to confirm AAR as the cause of the cracks.

Failed repair of a reinforced concrete beam

Refer to:
Concrete, Repair strategies, p. 14

Photo: *C. Stevenson, Cape Town, Republic of South Africa*

Sections of a harbour structure in the splash zone spalled severely as a result of extensive chloride-induced corrosion of the reinforcing steel. This set of photos illustrates the failure of a patch repair to one of the beams, because the preparation and patching procedure were inadequate.

The repair micro-concrete bonded only to the reinforcing steel bars and not to the residual concrete. The plaster applied to the sides of the beam hid this failure. A void between the repair concrete and the original beam concrete was revealed when the patch material was removed. It is

Photo: *C. Stevenson, Cape Town, Republic of South Africa*

Photo: *C. Stevenson, Cape Town, Republic of South Africa*

clear that the bottom reinforcing bars are no longer an integral part of the beam. It must be realised that chloride-induced corrosion is very pernicious and particularly difficult to repair successfully. Even if the exposed reinforcing steel is thoroughly cleaned and given a protective coating, corrosion could continue after repair if the surrounding concrete is still contaminated with chloride.

Metal failure

Corrosion of galvanised reinforcing steel

Refer to:
  Concrete, Repair strategies, p. 14
  Metal, Corrosion, p. 34
  Metal, Galvanised steel, p. 42

Photo: *J.R. Mackechnie, Cape Town, Republic of South Africa*

Bimetallic or galvanic corrosion occurs when two dissimilar metals are connected together in the presence of moisture. The more noble material is protected from corrosion at the expense of the other material, which in turn becomes anodic.

In general, galvanised reinforcement may be used together with plain reinforcing bars without any complications. It should be noted, however, that galvanised steel should not be considered as a replacement for reasonable cover depths and concrete quality. A floating concrete deck used in a marina structure was designed with galvanised reinforcement at only 20 mm cover in a moderate grade of concrete. Rapid ingress of chlorides soon depleted the zinc layer and resulted in serious corrosion damage of the structure within a few years of service. The first signs of damage were the appearance of longitudinal cracking immediately above the galvanised reinforcing bars.

Corrosion of a wrought-iron railing

Refer to:
  Metal, Suppression of metal corrosion, p. 37
  Metal, Restitution of corroded steel components, p. 41
  Paint, Causes of failure, p. 108

Photo: *R.O. Heckroodt, Cape Town, Republic of South Africa*

This wrought-iron railing of a European historic building shows both atmospheric and crevice corrosion, brought about by the breakdown of the protective paint layer. Obviously, the railing cannot be reinstated to its original condition (except by replacement), but the corrosion process can be considerably slowed down.

All traces of paint and corrosion products must be removed by shot or grit blasting. All crevices and lapping joints must be completely dry before the application of a rust-inhibiting primer, followed by the recommended number of coats of quality paints. If justified, a zinc or aluminium spray coating applied to the cleaned railing prior to painting would enhance the durability of the protection.

Corrosion of iron at setting points

Refer to:
   Metal, Corrosion, p. 34
   Metal, Suppression of metal corrosion p. 37

Photo: *R.O. Heckroodt, Cape Town, Republic of South Africa*

The localised galvanic corrosion of iron railings set in stone plinths using lead is a common occurrence. The usual remedy is to encase the iron (the anode) with a reliable coating in the vicinity of the joint, in order to prevent the formation of the corrosive species. This procedure is frequently ineffectual, because the concentration of all the galvanic current at breaks in the coating will result in localised but deep pitting. An alternative approach is to reduce the relative surface area of the cathode to that of the anode. This may be achieved by coating the cathode (the lead caulking) to reduce its effective surface area.

It is generally not possible to interrupt the ionic current return path, as the joint between the iron and lead cannot be protected completely from moisture, salt or polluting matter.

Red rusting of galvanised roof sheeting

Refer to:
   Metal, Corrosion, p. 34
   Paint, Causes of failure, p. 108

Photo: *H. Schollenberger, M&M Tek, CSIR, Republic of South Africa*

Red rusting of galvanised roof sheeting is generally the result of atmospheric corrosion in heavily polluted environments, due to the gradual depletion of the protective zinc layer. However, this example shows severe selective corrosion of a broiler house roof, although the chicken farm is situated in a rural area with no industrial pollution.

Over a period of 15 years, the roofing has developed extensive red rusting, concentrated towards the crest of the gable, adjacent to the ridge vent. The zinc coating is largely absent and general corrosion, as evidenced by red rust formation, has started. Observations revealed that the air blown out at the vent by forced draft ventilation is laden with dust particles containing organic phosphates (from animal dung) and chlorides (from chloride containing cleaning agents), as well as dissolved ammonia. The particles settle on the roof surface, particularly near the vent, where moisture (condensation of humid exhaust air, dew or light rain) leads to aqueous corrosion. Note that no corrosion of roof sheeting occurred above the vent, confirming that there was no general atmospheric corrosion due to environmental pollution. Investigations indicated that replacement of the roof sheeting was not necessary and that appropriate painting would prevent further corrosion.

In situations where red rust formation is localised and limited, the first and most important step is to mechanically clean the corroded steel down to bright metal. Abrasive paper or grit blasting must be used for this operation, because wire brushing is ineffective, as it merely polishes the red rust. The clean steel areas must be patch primed with a rust passivator, after which the whole area should be treated with a polyvinyl butyral based etch primer, followed by compatible topcoats. (See also the examples of painted galvanised steel (p. 124) and paint failure of galvanised roofing sheets (p. 127)).

Note that the top section of the roof was previously painted. However, this remedial action was not successful in the long run, as the coating failed after some time by flaking off, allowing the corrosion to continue. This failure is probably the consequence of poor preparation and inappropriate painting systems for this particular situation.

Where the red rust formation is extensive and the corrosive conditions severe, professional guidance is essential. In such situations industrial-type, heavy-duty primers and topcoats based on epoxy or polyurethane/acrylic based resins may be required to ensure an economic lifetime of the roofing system.

Deterioration of masonry

Irreversible moisture expansion of brickwork

Refer to:
Masonry, Differential movement, p. 53

This failure of a boundary wall (see opposite page) is similar in appearance to the example of the cracking apart of a brick foundation wall due to expansion of the concrete floor slab as a result of sulphate expansion (see the example of sulphate attack of a concrete slab (p. 120)). However, tests proved that sulphate attack was not a factor, but that this failure was caused mainly by the irreversible moisture expansion of the bricks. Repair of this unstable situation would involve rebuilding the corner section of the wall, incorporating efficient movement joints. Although much of the irreversible expansion has already taken place, allowance should be made for continued differential movement.

Photo: *reproduced, with permission, from R. Stubbs and K.E. Putterill (1972)* Expansion of brickwork, *Research Report 259. CSIR, Pretoria*

Differential movement in brickwork

Refer to:
   Masonry, Differential movement, p. 53
   Masonry, Chemical actions, p. 57

If brickwork is restrained from expanding, considerable forces can be built up as a result of differential movement, which can lead to crushing of the brickwork, as shown in (a) (see the photographs on the following pages). Balustrade and parapet walls, the ends of which are relatively free to expand, can become dangerously unstable if the expansive forces break down the mortar bond, particularly at the bed courses. Photograph (b) shows a parapet wall that has expanded to oversail the lower brickwork by up to 25 mm at the corners of the building. Tests proved that the major

(c)

Photos: *reproduced, with permission, from R. Stubbs and K.E. Putterill (1972) Expansion of brickwork, Research Report 259. CSIR, Pretoria*

contributor to the expansion that led to the two failures shown here was the irreversible moisture expansion of the bricks.

In most cases, damage due to differential movements can be avoided by applying correct design principles. Photograph (c) shows the failure of a face brick panel wall with a short offset, while Figure 7.1 illustrates the observed and recommended sections.

131

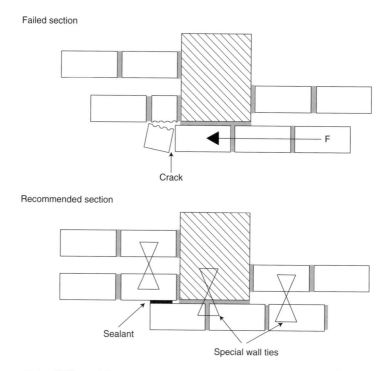

Figure 7.1 *Differential movement joints*

Lime blowing in concrete blocks

Refer to:
   Masonry, Chemical actions, p. 57

These craters (see opposite page) developed in a test wall constructed of concrete blocks. The wall was bagged (smeared or painted with a plaster slurry) and painted to show clearly the development of defects. The blocks were made with power station clinker ash that contained unslaked nodules of quick lime. The reaction between quick lime and water results in a considerable increase in volume, causing expansive disruption by the nodules. If the nodule is close to the surface, a small piece will pop out, forming a crater. Close inspection of the craters will reveal pockets of soft slaked lime at their base.

The craters shown in this plate developed within 6 months, but the pop-outs may occur for a number of years, depending on the temperature at which the coal has been burnt and the accessibility of moisture to the nodules. The structural integrity of the wall depends on the concentration of nodules in each block. If the incidence is low and the development rapid, repair could be delayed until the slaking process has been completed.

Photo: *J.E. Krüger, Pretoria, Republic of South Africa*

Salt action and efflorescence

Refer to:
Masonry, Chemical actions, p. 57

The photographs (see following page) show typical examples of the deterioration of low-fired clay bricks and pavers caused by the cryst-allisation of salts. The appearance of the failure is similar to that caused by frost action (see the example of frost attack (p. 137)), except that the presence of salts is very clear.

The salts causing the deterioration of the bricks were sulphates. The process is sometimes referred to as 'crypto-efflorescence', because the crystallisation of the sulphate salts during the drying cycle occurs mainly in, but not on, the surface layer of the brick. The influence of the degree of firing should be noted. The salts were sulphates from the coal used to fire the bricks and salts from sea spray.

133

Photo: *J.E. Krüger, Pretoria, Republic of South Africa*

Photo: *R.O. Heckroodt, Cape Town, Republic of South Africa*

The paving tiles are used as steps leading from a patio down to an estuary and they are in the splash zone. The salts were thus mainly chlorides, but sulphates may also be present. Note that the use of sealants or impregnants will increase the risk of further disintegration due to the continued crystallisation below the treated surface. Repair would require relaying the steps with hard-fired, low porosity tiles (vitrified, sometimes referred to as 'porcelain'), using an appropriate cement-based tile adhesive.

Chloride attack of calcium silicate bricks

Refer to:
   Masonry, Chemical actions, p. 57
   Masonry, Calcium silicate masonry, Durability, p. 66

Photo: *R.O. Heckroodt, Cape Town, Republic of South Africa*

135

This aspect of a modern house constructed with calcium silicate bricks is subjected to heavy salt-laden mists and wind-borne sea spray. Within 2 years of construction the smooth-faced bricks were severely eroded. Most of the bricks lost more than a centimetre and many of the mortar joints were standing proud. Protected aspects of the building showed only slight deterioration, while the unrendered interior walls of the motor garage were unaffected.

The friability of the bricks was caused by the breakdown of the calcium silicate matrix by chlorides. Attempts to prevent the ingress of salt by sealers and impregnants were unsuccessful. It is unlikely that rendering the exposed walls will be a successful repair, because any residual salt in the porous bricks will continue their breakdown process.

The poor resistance to frost attack of salt-contaminated calcium silicate bricks is the result of the same breakdown mechanism.

Sulphate attack of mortar

Photo: *R.O. Heckroodt, Cape Town, Republic of South Africa*

Refer to:
  Masonry, Chemical actions, p. 57
  Concrete, Chemical attack on concrete, p. 25

The pronounced bending of this chimney stack of a brandy distillery (see opposite page) was caused by the preferential soaking of the brickwork by heavy rain driven from one direction. This led to enhanced sulphate attack of the mortar on that side of the stack, resulting in preferential expansion.

Because demolition is not an acceptable action for this landmark, the structure must be stabilised, preferably by inserting an armature.

Frost attack

Refer to:
  Masonry, Environmental impact, p. 55

Photo: *R.O. Heckroodt, Cape Town, Republic of South Africa*

The disintegration of these pavers caused by frost action is similar to that caused by crystallisation of salts (see the example of salt action and efflorescence (p. 133)), and therefore tests for soluble salts should be done.

The selective deterioration of some of the pavers emphasises the critical influence of small differences in pore structure on the resistance to frost attack. Only pavers with a proven track record of resistance to frost action should be used in these service conditions.

Tiling failure

Failure due to exceeding the open time of the adhesive

Refer to:
  Tiling, Tile adhesive failure, p. 78

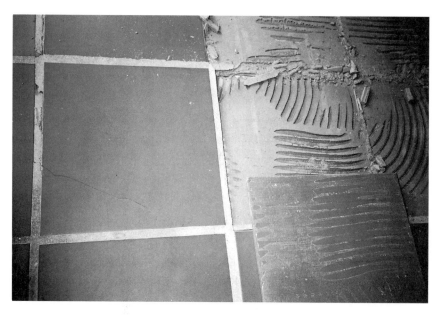

Photo: *R.O. Heckroodt, Cape Town, Republic of South Africa*

The tiling of a domestic outdoor patio in a temperate geographical region failed in adhesion after about 3 years of service. The tiles were semi-vitreous and 300 mm by 300 mm in size. The moisture expansion of the tiles was relatively low (<0.03%). The maximum daily surface temperature variation was about 40°C, which would have resulted in a maximum thermal expansion or contraction of about 0.025%.

Most of the tiles de-bonded over a relatively short period after laying, but there was little sign of doming or ridging. De-bonding occurred only at the tile–adhesive interface, never at the adhesive–screed interface. As can be seen from the photograph, a fair amount of contact between tile and adhesive was achieved during the laying operation. However, only small amounts of adhesive remained stuck to the tiles at failure.

It would seem that the open-time limit of the adhesive was exceeded during the tiling operation. After applying the adhesive with the appropriate notched trowel to the screed, the tiles were placed and bedded when the adhesive had already dried sufficiently on its surface to prevent effective wetting of the tile surface. This situation could arise when the adhesive is applied to an area that is too large, when the placing of the tiles is delayed, or when the ambient conditions are such that the surface of the applied adhesive dries rapidly.

Failure of external tile cladding

Refer to:
  Tiling, Differential movement, pp. 82, 84

Photo: *J.E. Krüger, Pretoria, Republic of South Africa*

The main cause of this tiling adhesion failure is the combined effect of irreversible moisture expansion of the glazed split tiles and the drying shrinkage of the concrete wall.

The wall was clad with ceramic split tiles, which offered a good mechanical key between the adhesive (sand–cement mortar) and the tiles. The joints were grouted with a sand–cement mortar. Failure occurred between the relatively smooth concrete and the mortar, about a year after cladding of the wall. The tiles had a moisture expansion of about 0.1%, which placed them in the high expansion category.

Movement joints were not provided as recommended, but even if they were, it is very unlikely that they would have prevented the failure.

Tiling failure caused by differential movement

Refer to:
Tiling, Differential movement, pp. 82, 84

Photo: *J.E. Krüger, Pretoria, Republic of South Africa*

This adhesion failure of patio tiling occurred about 2 years after construction. The punch tiles measure 300 mm by 300 mm and their water absorption was in the range 4–6%. The irreversible moisture expansion of the tiles as received was 0.15%, which places them in the high expansion category. The residual (still to be expected) expansion of the exposed tiles was determined as 0.07%, and therefore the patio tiles had expanded by about 0.08% when failure occurred.

There was no information on the amount of drying shrinkage that occurred in the concrete base slab. However, the tiles were apparently laid days rather than weeks after the concrete was placed. Under such conditions the shrinkage of concrete can be 0.03–0.05%.

The photograph also clearly shows that the tiles were poorly bedded and that there was little adhesion between the tiles and adhesive. The failure was thus the combined effect of differential movement between the expanding tiles and shrinking concrete base, as well as poor workmanship. Note that the use of so-called 'expansion joints' would not have avoided this type of failure.

Lime blowing in floor tiles

Refer to:
   Tiling, Lime blowing, p. 76
   Masonry, Lime blowing, p. 57

Photo: *R.O. Heckroodt, Cape Town, Republic of South Africa*

The photograph shows one of the larger craters caused by lime blowing, that appeared about 2 years after the tiles had been laid (the diameter of the coin is 15 mm). The tiles had a porosity of less than 1% and they were glazed with a thin vitrifying engobe.

Less than 3% of the tiles developed craters larger than about 5 mm, but only a few of them were readily noticeable, because of the similar colour of the body and the surface layer. When lime blowing occurs in dark-coloured tiles with a light-coloured glazed surface, and vice versa, even small craters are very noticeable.

Tile manufacturers can largely avoid this fault if they grind the raw materials that contain lumps of limestone or dolomite finer.

Timber deterioration

Wet rot of pergola timbers

Refer to:
  Timber, Fungi (decay), p. 88

Photo: *R.O. Heckroodt, Cape Town, Republic of South Africa*

The connection between the two untreated, non-durable hardwood (meranti) beams of this pergola shows serious localised wet rot attack. When exposed, the attacked wood around the joint is characteristically fibrous, light coloured and soft. The surface coating appears mainly intact. However, wide gaps in the coating where the two beams make contact allow easy penetration of rainwater or dew. Moisture remains trapped inside the coated wood much longer, creating ideal conditions for fungal attack. Small, biscuit-shaped fruit bodies have developed on top of the lower beam. Fruit bodies are also visible near the end grain of the upper beam.

These are typical design errors, as fully exposed lap joints and end-grain regions in untreated structures can be particularly susceptible to fungal attack. Paints or varnishes are not effective in stopping moisture penetration at such positions. Although water-repellent penetrating stains require more frequent maintenance than paints and varnishes, they should be able to exercise better moisture control. Replacement of affected sections only is frequently not feasible and total replacement with pretreated timber is then the alternative.

Wet rot of a hardwood door

Refer to:
   Timber, Fungi (decay), p. 88

Photo: *T. Rypstra, Stellenbosch, Republic of South Africa*

A very common example of wet rot attack is shown at the bottom of the vertical style of this hardwood (meranti) door. During the construction and finishing phase, the end grain at the bottom of the style had not been properly sealed or coated. Rainwater entered the wooden style by capillary movement during exposure to weathering. The coating could not absorb the dimensional changes associated with the increased differential swelling and shrinkage of the vertical style and horizontal bottom section of the door, resulting in cracks in the coating over the joint. Regular penetration of water occurred through the damaged wood finish over the failed joint. Similar defects can occur in the lower sections of window frames.

Specialised repair systems are available. Replacement of wooden door and window joinery elements is frequently easier but more costly.

Wood-disfiguring fungi

Refer to:
Timber, Fungi (decay), p. 88

Photo: *G.C. Scheepers, Stellenbosch, Republic of South Africa*

The transparent varnish on this pine sample is severely weathered and blue-stain fungi (commonly *Aureobasidium pullulans*) have developed on the wood surface below the varnish coating. Further exposure to outdoor weather may cause deeper discoloration. Although not destructive, the presence of these wood-disfiguring fungi is a serious warning that the conditions are favourable for attack from other organisms such as the wood-destroying fungi and specific insects. The varnish must be stripped, surfaces sanded and re-coated to correct this defect.

Blue-stain fungi can also attack timber inside buildings. Their presence in wooden floorboards, wall cladding, ceilings, etc., indicate that water is leaking into or accumulating in the structure and, if not attended to, may lead to development of dry and/or wet rot. Important remedial steps to be taken are that the source of the moisture must be eliminated and adequate ventilation must be established without delay. Once established, deep blue-stain discoloration (e.g. in decorative wooden products) cannot be removed.

Be aware that localised green or dark discoloration may develop as a result of chromated copper arsenate (CCA) or other copper containing treatments of timber. Sapwood is more permeable than heartwood and, during CCA or other preservative treatment, higher retention and penetration are obtained in the sapwood regions. Exposure of these freshly treated timber surfaces to direct sunlight further increases the intensity of the discoloration, while varnishes generally enhance these colour differences in wood even more.

Insect attack: Cerambycid borers

Refer to:
  Timber, Insects, p. 94

Photo: *T. Rypstra, Stellenbosch, Republic of South Africa*

An untreated, softwood (*Pinus pinaster*) pole section of a truss of a rustic thatch roof is under active attack by the European house borer (*Hylotrupes bajulus*). A few large oval flight holes have appeared, but they are widely spaced and not easily detected. However, frass is visible as white spots on the rough, brown-yellow surfaces of the pole. The hardwood (a *Eucalyptus* species) battens are not affected, as this insect does not attack this hardwood species. The number of flight holes is not a good indication of the extent of damage, as a widespread tunnelling network inside could have already seriously compromised the structural integrity of the pole.

Remedial treatment can consist of heat treatment, fumigation or drilling holes and injecting a preservative in them, followed by spraying. Drilling also helps to assess whether replacement of roof truss sections or complete trusses may be necessary.

Insect attack: Anobiid borers

Refer to:
Timber, Insects, p. 94

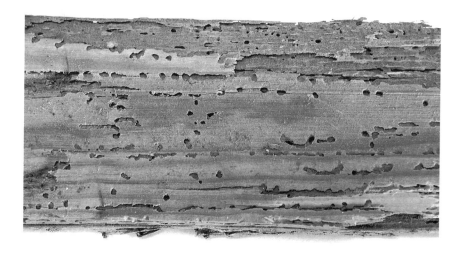

Photo: *G.C. Scheepers, Stellenbosch, Republic of South Africa*

Attack by the common furniture beetle (*Anobium punctatum*), shown here in a section through a piece of softwood floorboard, is well advanced and dangerous. Attack in timber floors can be more severe if subfloor ventilation is poor and humid conditions underneath the floor are maintained. Vinyl floor covering over wooden floorboards, poorly planned alterations to buildings and poor maintenance often result in increased risk of attack by this woodborer. Round flight holes and coarse granular frass lying in small heaps over flight holes are characteristic indicators. As the name implies, these borers can typically be found in old furniture, a likely source of spreading infestation.

Attack of floorboards requires treatment of all subfloor timber. Fumigation, 'drill and inject' of floor beams and/or flooding or replacement of structurally weakened boards can be considered as remedial actions but can only be undertaken by specialists.

Powder post termite

Refer to:
  Timber, Insects, p. 94

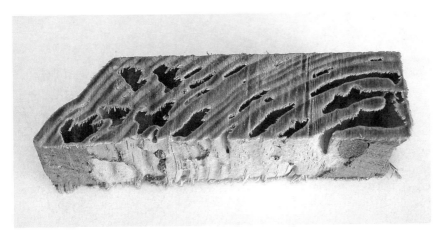

Photo: *G.C. Scheepers, Stellenbosch, Republic of South Africa*

A cross-section of this softwood cupboard component clearly reveals the characteristic extensive tunnelling by powder post termites. The soft early wood (springwood) has been completely consumed, leaving the harder late wood (summerwood) largely untouched. Faecal pellets resembling poppy seed may be found inside galleries. Damage goes largely unnoticed until a few emergence holes, associated with advanced infestation, appear.

Complete fumigation of the building is recommended.

147

Marine borers

Refer to:
  Timber, Marine borers, p. 94

Photo: *R.O. Heckroodt, Cape Town, Republic of South Africa*

The relative resistance of creosoted softwood piles and sawn softwood cross-bracing members of this pier against attack by marine borers (*Limnoria*) (and probably fungi as well) are clearly demonstrated. The sawn timber cross-bracing members were not adequately treated and some were completely destroyed in the fluctuating tidal zone in which *Limnoria* were active, and favourable conditions for fungal growth were maintained. The outer sapwood 'shell' of the piles was pressure treated adequately, meaning that sufficient creosote was deposited deep enough to control attack. In sawn-timber building elements such as cross-bracing members, the exposed heartwood sections are often difficult to treat, resulting in reduced performance. Deep penetration and high retention of preservatives are essential for all timber used in marine environments in order to give maximum protection. All machining, such as drilling and sawing, to boards or poles must be completed before treatment to avoid exposure of sections that were difficult to pressure treat to adequate retentions of preservative.

Moisture damage

Refer to:
  Timber, Weathering, p. 96

Photo: *T. Rypstra, Stellenbosch, Republic of South Africa*

Water spillage on this hardwood parquet floor underneath and around the flowerpot has resulted in swelling, adhesion failure and staining of the wooden blocks.

Remedial repairs and restoration are more successful if immediate action is taken. The process is time consuming. Liquid water must be mopped up, and all loose and affected blocks must be removed and air dried immediately. The underlying concrete floor should be allowed to dry out completely and the blocks refitted or, when seriously stained, replaced. Floor sanding might be necessary to correct surface irregularities resulting from warping and cupping of the blocks or the adhesive between the concrete and the blocks.

Paint failure

Flaking from new cementitious substrate

Refer to:
  Paint, Concrete and other cement products, p. 112

Photo: *M.S. Smit, Boutek, CSIR, Republic of South Africa*

The dramatic flaking of the paint from a new cementitious-based ceiling material occurred because of improper preparation of the substrate. The paint system was applied directly to a new surface containing efflorescing salts formed during the drying of the cementitious ceiling material.

The repair requires the complete removal of all the layers of failed paint, as well as all the efflorescing salts, using only a dry operation. Repainting should be done with 'contract' emulsion paint, adhering strictly to the method that the manufacturer recommends for ceilings.

Delamination of newly painted surfaces

Refer to:
  Paint, Instability of the substrate, p. 108
  Paint, Improper preparation of substrate, p. 108

Photo: *M.S. Smit, Boutek, CSIR, Republic of South Africa*

Failure of the paint is attributed to improper preparation of the previously painted surface and some instability of the substrate. Weathering of some paints usually results in chalking on the surface. If this powdery layer is not properly removed and the substrate repaired, delamination of superimposed new paint layers will occur.

The failed paint and chalking should be removed and the substrate repaired. Apply a base coat of a pure acrylic emulsion paint thinned down, followed by one, or preferably two, coats of good-quality pure acrylic emulsion paint.

Staining by micro-organism growth

Refer to:
Paint, Improper preparation of substrate, p. 108

Staining of the paint on the ceiling (see following page) occurred as a result of micro-organism growth. The ceiling was poorly insulated in the areas where condensation and subsequent micro-organism growth occurred.

Repairs will entail the correct installation of insulation in the ceiling, followed by scraping and brushing off the stains in the dry condition. The ceiling, and especially the stained areas, must be treated with a diluted sodium hypochlorite solution (diluted household disinfectant). Time

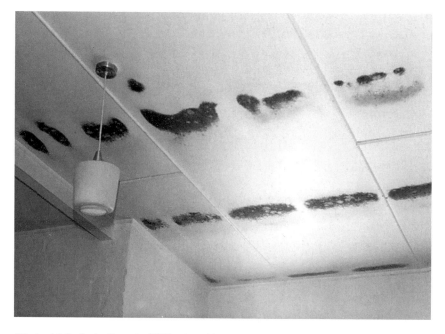

Photo: M.S. Smit, Boutek, CSIR, Republic of South Africa

should be allowed for the proper drying of the ceiling materials, and then one or two coats of an emulsion-based ceiling paint incorporating a biocide applied.

Paint failure on an exterior hardwood door

Refer to:
Paint, Moisture in substrates, p. 113
Timber, Weathered timber, p. 100

The paint failed (see opposite page) because of ageing, because its adhesion to the wood was destroyed by water in the wood being drawn to the surface or because changes in moisture content caused excessive movement of the wood. The paint film may also have failed by loss of adhesion to the wood because a poor-quality primer was used. This type of paint failure is often a combination of all these reasons.

To repair the door, strip the wood of all paint. Remove rusting fasteners and replace with non-rusting items. Repair all damaged and deteriorated wood. Re-prime with a solvent-borne wood primer and repaint with a pure acrylic based topcoat(s).

Photo: *T. Rypstra, Stellenbosch, Republic of South Africa*

Redecorating a garden gate

Refer to:
   Paint, Metals, p. 109
   Paint, Timber, p. 112

The deterioration of the garden gate (see following page) is the result of poor materials selection and bad workmanship. The woodwork was redecorated badly and without proper preparation of the old paint surfaces, while gate fittings not suitable for exterior exposure were used. The paint was not able to endure the dimensional movements in the wood, caused by the periodic ingress of moisture through cracks and gaps in the paintwork.

In order to repair the paintwork, the gate fittings should be removed and replaced with galvanised metal fittings. The stained paint and affected paint should be completely scraped and/or abraded off. The wood should be primed with a solvent-borne wood primer and painted with a pure acrylic-based topcoat(s). If required, paint the galvanised fittings with a polyvinyl butyral based etch primer and finish off with compatible topcoat(s).

Photo: *R.O. Heckroodt, Cape Town, Republic of South Africa*

Painted galvanised steel

Refer to:
Paint, Improper preparation of the substrate, p. 108
Metal, Galvanised steel, p. 42

Galvanised surfaces require correct preparation before painting. The peeling of the paint on this galvanised support pole of a carport (see opposite page) may be attributed to the use of an incompatible primer or even lack of primer and/or to improper surface preparation. Before repainting, scrape off all the loose paint, followed by light sanding. Paint the clean substrate with a polyvinyl butyral based etch primer and finish off with compatible topcoat(s).

The corrosion on the underside of the galvanised sheeting is 'white rust' formation on the zinc surface caused by repeated condensation and evaporation as a result of diurnal climatic changes. The first remedial step is to use water to wash and brush the zinc oxide off. Carefully inspect the cleaned galvanised surface for red rust. If no red rusting has occurred, apply a polyvinyl butyral based etch primer and compatible topcoat(s). If red rust did occur, abrade the rust spots to the clean metal state and treat with a rust passivator. Follow up with a polyvinyl butyral based etch primer and compatible topcoat(s), as specified for galvanised surfaces.

Photo: *R.O. Heckroodt, Cape Town, Republic of South Africa*

Paint failure of galvanised roofing sheets

Refer to:
Paint, Metals, p. 109

The paint failure in the photograph (see following page) is attributed to improper preparation of the galvanised steel substrate. Galvanised zinc surfaces are always factory treated to prevent storage staining (white rust) during wet storage. Galvanised sheet surfaces thus need careful preparation before pretreatment or painting, otherwise peeling and flaking of paints will occur readily.

To repair the paintwork, all paint should be removed and the zinc surface should be thoroughly cleaned by light abrasive cleaning followed by detergent washing and rinsing with water. Repaint with a polyvinyl

Photo: *M. S. Smit, Boutek, CSIR, Republic of South Africa*

butyral based etch primer followed by compatible topcoats (see also the example of red rusting of galvanised roof sheeting (p. 127)).

Bibliography

There are numerous publications dealing with building materials, either in general or with emphasis on specific aspects of their deterioration and failure. Some of the more recent publications are listed here as a general guide to the available literature.

Construction materials

Doran D.K. 1994. *Newnes construction materials pocket book*. Newnes, Oxford.
Everett A. 1994. *Materials – Mitchell's building series*, Longman Scientific & Technical, Harlow.
Taylor G.D. 1991. *Construction materials*. Longman Scientific & Technical, Harlow.
van Amsterdam E. 2000. *Construction materials for civil engineering*. Juta, Cape Town.
Young J.F., Mindess S., Gray R.J. and Bentur A. 1998. *The science and technology of civil engineering materials*. Prentice Hill, Englewood Cliffs, NJ.

Building defects

Cook G.K. and Hinks A.J. 1992. *Appraising building defects*. Longman Scientific & Technical, Harlow.
Eldridge H.J. 1976. *Common defects in buildings*. HMSO, London.
Hinks J. and Cook G. 1997. *Technology of building defects*. E. & F.N. Spon, London.
National Building Agency 1983. *Common building defects*, Longman Scientific & Technical, Harlow.
Ransom W.H. 1981. *Building failures*. E. & F.N. Spon, London.
Richardson B.A. 1991. *Defects and deterioration in buildings*. E. & F.N. Spon/ Chapman and Hall, London.
Weaver M.E. 1993. *Conserving buildings*. Wiley, New York.

Standards and codes of practice

National and regional standards, specifications and codes of practice should be constantly checked for updated versions.

Web sites

There are a large number of web sites covering topics dealing with the deterioration and failure of building materials and associated subjects. Many of these sites are updated regularly, while new ones are continually being created. Most of the web sites are commercial in nature.

Some entry-point web sites are listed below. Every effort has been made to ensure the correctness of the internet addresses, but no responsibility is assumed for any errors or omissions.

American Society for Testing and Materials: http://www.astm.org

Building Research Association, New Zealand: http://www.branz.org.nz

Building Research Establishment, UK: http://www.bre.co.uk

Construction Industry Acronyms List: http://www.ciob.org.uk

Division for Building, Construction and Engineering, CSIRO, Australia: http://dbce.csiro.au

Division of Building and Construction Technologies, CSIR, South Africa: http://www.csir.co.za

Entry website by the NIST (National Institute of Standards and Technology) to National and International Standards Organizations: http://www.nist.gov/oiaa/stnd.org.htm

Publisher's list: http://www.cnu.edu/library/aurls.htm

Index

Note: Page references in *italics* refer to figures; those in **bold** refer to tables

acrylic resin-based etch primers 109
alkali–aggregate reaction of concrete
    22–5, 121–2
  conditions necessary for 24
  confirmation of 23–4
  environmental conditions 24
  high alkalinity of pore solution 24
  preventive measures 24
  reaction active 25
  reaction dormant 25
  reactive phases in the aggregate 24
  recognition of 22–3
  remedial action 25
  types 22
alkali–carbonate rock reaction (ACR)
    22
alkali–silica reaction (ASR) 22
aluminium 46–9
  age (precipitaion) hardening 46
  alloying 46
  anodised 48–9
  atmospheric corrosion 47
  corrosion at concrete interface 47–8
  crevice corrosion 48
  galvanic corrosion 47
Anobiid borers 102, 146–7
Anobium punctatum (common
    furniture beetle) 146–7
Aspergillus spp. (mould) **90**
atmospheric corrosion
  of galvanised steel 42
  of metal 35
Aureobasidium pullulans (blue stain
    fungi) 144

bacteria, timber deterioration and 96
  remedial action **99**

beetles (Coleoptera), timber
    deterioration and **92**, 94, **95**
  remedial action **99**
bimetallic corrosion of metal *see*
    galvanic corrosion of metal
biocides, use of 99–100
biological agents 88
black rust 39
blue stain fungi **94** 144
Bostrychid spp. (wood borers) 88
brick burn 69
brown rust 39

calcareous sandstones 65
calcium silicate masonry 66–7
  differential movement 67
  durability 66–7
  frost action 67
calcium zincate formation 43
carbonation depth of concrete 12, *12*
carbonation-induced corrosion of
    concrete **18**, 119–20
carbonation shrinkage of concrete 30
cathodic protection (CP) systems 21, *21*
cellar fungus **93**
cementitious substrate, paint flaking
    from 150
Cerambycid borers 145–6
ceramics, glazed, crazing of 76
chemicals, timber deterioration and 98
chloride attack of masonry 59, 135–6
chloride-induced corrosion of
    concrete
    carbonation-induced corrosion cf.
    18
    migrating corrosion inhibitors 19,
    19, 25

patch repair 15
qualitative risk *13*
chloride-induced corrosion of steel 40,
    122–34
chloride testing of concrete 7, *11*,
    118–19
    qualitative risk of corrosion based
        on chloride levels *11*
chromated copper arsenate (CCA) 145
Cladosporium spp. (mould) **90**
coherent layers 37
concentration cells 36
concrete
    alkali–aggregate reaction 22–5,
        121–2
    bricks and blocks 73–4
    carbonation-induced corrosion *see*
        chloride-induced corrosion **18**,
        119–20
    chemical attack 25–7
    chloride-induced corrosion *13*, 15,
        18, 19, *19*, 25, 118–19
    conditions and features **4–5**
    corrosion damage 2–22
    demolition or reconstruction 22
    diagnostics of deterioration **3**
    dimensional change 29–31
    fire damage 28
    frost attack 28–9
    manifestations of corrosion
        damage 2–6
    mortar and plaster 60–1, **62–3**
    paint failure 112
    reinforcement corrosion, condition
        surveys of 6, 7–14, **8–9**
    repair strategies 6, 14–22, **16–17**, 25,
        122–4
    sulphate attack of a concrete slab
        120–1
    three-stage model of corrosion
        damage *6*
condition surveys of reinforcement
    corrosion 6, 7–14
Coniophora puteana (wet rot) **90**, **93**
contour scaling of natural stone
    masonry 65
copper alloys 49–50, **49**
    atmospheric corrosion 50
    dezincification of brass 50

galvanic corrosion 50
copper–zinc alloys (brass) 49
    dezincification of 50
corrosion, metal 34–8
corrosion rate measurements of
    concrete 14, **14**
corrosion-resistant steel (weathering
    steel) 44
corrosion threshold value of
    reinforcing steel 40
cracking of concrete, alkali–aggregate
    reaction and 23
crazing of glazed ceramics 76
creep
    of cement substrate, and tiling
        failure 83
    of concrete 30
    of masonry 55
crevice corrosion of metal 36
    suppression of 38
crypto-efflorescence 60, 133
Cryptotermes brevis (powder post
    termite) **91**
crystallisation of salts 59–60, 65–6,
    133–5

dampwood termites **91**
death watch beetle **90**
dedolomitisation 22
delamination of newly painted
    surfaces 150–1
dimensional change of masonry 55
discoloration of concrete, alkali–
    aggregate reaction 23
drying shrinkage of concrete 30, 83

efflorescence of masonry 58–9, 133–5
electrochemical chloride removal
    (ECR) 19–21, *20*
electrochemical repair techniques of
    concrete 19–21
erosion corrosion of metal 37
European house borer **92**, 146
expansion joints in structures 141
expansion of concrete members, alkali–
    aggregate reaction 23

fire damage
    of concrete 28

of masonry 56–7
    remedial action 28
    of timber 100
fired clay masonry 67–73
    attack by salts 69
    cleaning clay brickwork 69–73, **70–2**
    differential movement 67–8
    frost resistance 68–9
    lime blowing 68
frost attack
    of calcium silicate masonry 67
    of concrete 28–9
    of masonry 55–6
    of natural stone masonry 65, 137–8
    of tiles 78
frost resistance of paving and fired
        clay masonry 68–9
fungi (decay), timber deterioration
        and 88–94
    conditions necessary for
        development 88
    identification of biological attack
        90–2
    life cycle of fungi 89
    remedial action **99**, 100
    wood-destroying (rot or decay) 89,
        93, 142–3, 143–4
    wood-disfiguring 89, **94**, 144–5

galvanic (bimetallic) corrosion of
        metal 35–6, 124–5
    galvanised steel 42
    suppression of 38, **38**
galvanised steel 42–3, 124–5
    atmospheric corrosion 42
    coatings for 109
    paint failure 154, 155–6
gel, presence of, alkali–aggregate
        reaction 23
gypsum plasters 61, 112
gypsum skim coats 112

hardwoods 102
    paint failure on 152
    wet rot 143–4
heat
    concrete and 28
    timber deterioration and 98
Hylotrupes bajulus (European house
        borer) **92**, 146
hyphae, fungal 89

impressed current cathodic protection
        (CP) systems 21
incoherent layers 37
insecticides 100, 102
insects, timber deterioration and 94,
        145–6, 146–7
    remedial action **99**, 100
    resistance to attack 102
intergranular corrosion of metal 37
iron, corrosion at setting points 126–7
irreversible moisture expansion
    of masonry 54–5, 128
    of tiles 82–3, **83**

lime blowing
    of fired clay masonry 68
    of masonry 57–60, 132
    of paving 68
    of tiles 76–7, 141–2
limestone 57
Limnoria (marine borers) **97**
linear polarisation resistance (LPR)
        principles 14
Lyctid beetles 102
Lyctus brunneus (powder post beetle) **91**

macro-cell corrosion of reinforcing
        steel 40
manganese dioxide in masonry 58
map cracking of concrete 121–2
marine borers, timber deterioration
        and 94–6, **97**, 148–9
    remedial action **99**
masonry, deterioration of 128–38
    causes of 52–60
    chemical actions 57–60, 129–31,
        133–7
    differential movement 53–5, 128–32
    environmental impact 55–7
    moisture 52–3
mechanical forces, timber
        deterioration and 98–9
    remedial action 100
metal failure 34–7, 124–8
    corrosion 34–7
    mechanical failure 34

passivation 37
protection of exposed metal
    surfaces 37
suppression of metal corrosion 37–8
metals, paint failure on 109–12, 153
micro-cell corrosion 40
micro-organism staining of paint
    151–2
moisture damage
    masonry 52–3
    paint failure and 113–14
    timber 88, 149
    see also irreversible moisture
        expansion
mine fungus **93**
mortar 60, **62–3**
mould 89, **90**, **94**
mycelium, fungal 89

natural absorption of water 52
natural stone masonry 61–6
    contour scaling 65
    crystallisation of salts 65–6
    durability 61, **64**
    frost attack 65, 137–8
    surface appearance 61–5

paint failure 107–8, 150–6
    causes 108–14
    common **110–11**
    on metals 37, 153
    moisture in substrates 113–14
    preventive and remedial actions
        114–15
    quality of paint 114
    on steel 41–2
    substrate instability 108
    substrate preparation, improper
        108–13
    unsuitable paints 113
patch repairs of concrete 15
    chloride-induced corrosion 15
    incipient anode formation after *18*
paving
    attack by salts 69
    differential movement 67–8
    frost resistance 68–9
    lime blowing 68
Penicillium spp. (mould) **90**

physicochemical agents 88
pitting corrosion of metal 36
plaster 60, **62–3**
plaster of Paris 61
pollution, masonry and 55
polyvinyl butyral 109
porosity 52
potassium carbonate efflorescence
    58–9
potential map, rebar potentials 12
powder post beetle **91**
powder post termite **91**
preservatives, chemical, of timber 102,
    103, 104

re-alkalisation of concrete 19
rebar potentials of concrete 12–13, **13**
reinforcement corrosion of concrete,
    condition surveys of 6, 7–14,
    8–9
rendering 60
repair strategies of concrete corrosion
    damage 6, 14–22, **16–17**
resistivity of concrete 13–14, **14**
rust
    red 39, 127–8, 154
    brown 39
    black 39
    white 109, 154

sacrificial anode systems 21
salt attack of tiles 78
sap stain (blue stain) **90**
sapwood 102
saturation coefficient 52
screeds 60
Serpula lacrymans (dry rot fungus) **93**
shrinkage
    of concrete 30
    of masonry 55
siliceous sandstones 65
soft rots 89, **93**
soft water, concrete and 26
    attack mechanism 26
    remedial action 26
softwoods 102
spores, fungal 89
sporophores, fungal 89
staining fungi 89, **94**

stainless steel 44–6
  austenitic 44, **45**
  cleaning 45, **46**
  ferritic 44, **45**
  painting 46, 109
  repassivation 45
stress corrosion of metal 36–7
  suppression of 38
structural steel 38–42
  atmospheric corrosion 39
  corrosion of steel embedded in
     concrete 39–41, **41**
  depassivation 39–40
  restitution of corroded steel
     components 41–2
sulphate attack
  of concrete 26–7, 120–1
  of masonry 59, 136–7

tarnishing of metal 35
Teredo navalis (marine borer) **97**
termites (Isoptera), timber
     deterioration and 94, **96**, 97,
     147
  remedial action **99**
termiticides 100
thermal expansion
  of concrete 30–1
  of masonry 53, **53**
thermal movement
  of masonry 53
  of tiles 82, **82**
tile adhesive failure 78–80
  attack by aggressive chemicals 80
  exceeding open time 138–9
  inappropriate adhesive 79
  poor workmanship 80
tile adhesives
  class classification **79**
  generic classification **79**
tiled floors and walls 81–5
  differential movement 82–3, 84,
     1401
  failure of tiled systems 83–4

movement joints 85
  preventive measures 84–5
  workmanship 85
tiling failure 138–42
  external tile cladding 139–40
  floors and walls 81–5
  material failure 76–81
timber deterioration 142–9
  agents of attack 88–99
  biological attack 99–100, **99**
  environmental control 104
  moisture damage 149
  paint failure 112, 153
  preventive actions 101–4, **101**
  providing barriers 104
  remedial actions 99–101
  resistance to decay 102
  selection of timber 102
total absorption of water 52

vanadium efflorescence 58
visual assessments of concrete 7, **10**

wash primers 109
weathering steel (corrosion-resistant
     steel) 44
weathering, timber deterioration and
     96–8, 100–1
wet rots 89, **93**
white rot fungus **93**
white rust 109, 154
woodborers 102
wood disfiguring fungi 89
woodworm 95
wrought-iron railing, corrosion of
     125–6

Xestobium rufovillosum (death watch
     beetle) **90**

zinc coating of steel 42–3, **43**, 127–8
zinc primers 109
Zootermopsis spp. (dampwood
     termites) **91**